THE ROLLS-ROYCE SPEY

Michael Hartley

TECHNICAL SERIES No 10

Published in 2008 by
The Rolls-Royce Heritage Trust
PO Box 31, Derby, England DE24 8BJ

ISBN: 978-1-872922-26-3

The Historical Series is published as a joint initiative by
The Rolls-Royce Heritage Trust and the Sir Henry Royce Memorial Foundation.

Previous volumes published in the Series are listed at the rear, together with
volumes available in The Rolls-Royce Heritage Trust Technical Series.

Cover picture: Colour photo of RB168-25R on test bed, reheat lit.

Books are available from:
The Rolls-Royce Heritage Trust, PO Box 31, Derby DE24 8BJ

Printed by: blp (Northern) Ltd.

CONTENTS

FOREWORD

Michael Hartley

This book is a memorial to its author – Michael Hartley. When Michael died after a lifetime's struggle with ill-health (though unless one knew him closely it was never obvious), I was asked by the Heritage Trust to complete what Michael had already written, which in fact turned out to be the majority of the main body of the book. My task has been to add the chapter on the Spey in Scotland, and most of the Appendix and Afterthoughts, though strictly under Michael's influence as he left copious notes on what was to be included. I also looked out the illustrations, but again following Michael's instructions.

It is typical of everything that Michael did in his career at Rolls-Royce that this is a painstakingly researched work full of detail on the most widespread family of all the Rolls-Royce gas turbines. The complexity of the Spey family is such that Michael felt the story should not include the Tay, even though it is a very worthy offspring, as that engine merits a book in its own right.

Those of us who were privileged to know Michael well will recognise his unique wit throughout the book, and particularly in one section in the Afterthoughts. In fact, the whole text bristles with his down-to-earth style and his love of the nicknames that Rolls-Royce engineers gave to particular features of the engines and his love of a good story brings life to the book in many places.

I do not have a list of all those whom Michael consulted in the preparation of this book, but I know he spoke to Don McLean, and many of his design team, as well as Giles Harvey and Chris Webber and members of their development team. He was particularly pleased to have help from Bill Castle and Jerry Roan of Allison. Dave Pratley supplied information from Ansty – on the day before his retirement from Rolls-Royce! Fred Steele of McDonnell supplied his thoughts which I know were much appreciated by Michael. To those whose names I have not mentioned, my apologies but nevertheless grateful thanks. Among the many people who have assisted me I particularly wish to thank Giles Harvey, Colin McChesney, Gerald Davies, Keith Hatchett, Ken Goddard, Wilf Ewbank and Cyril Elliott for their help.

There are not many people of a certain age at Rolls-Royce whose lives were not touched at some point by the Spey in one form or another. It was a great influence on the career of many of us and it was also a tremendous influence on the development of Rolls-Royce in all of the various fields in which gas turbines are used. Among the many benefits to the Company, it gained us a foothold in both the civil and military markets of North America, both of which were later exploited by the RB211, the Adour and the Pegasus.

In reading this remarkable book about a defining period in the history of Rolls-Royce, it is impossible not to be aware of the character of its author – an intelligent, humourous, scrupulously honest, modest engineer – Michael Hartley.

Alec Collins

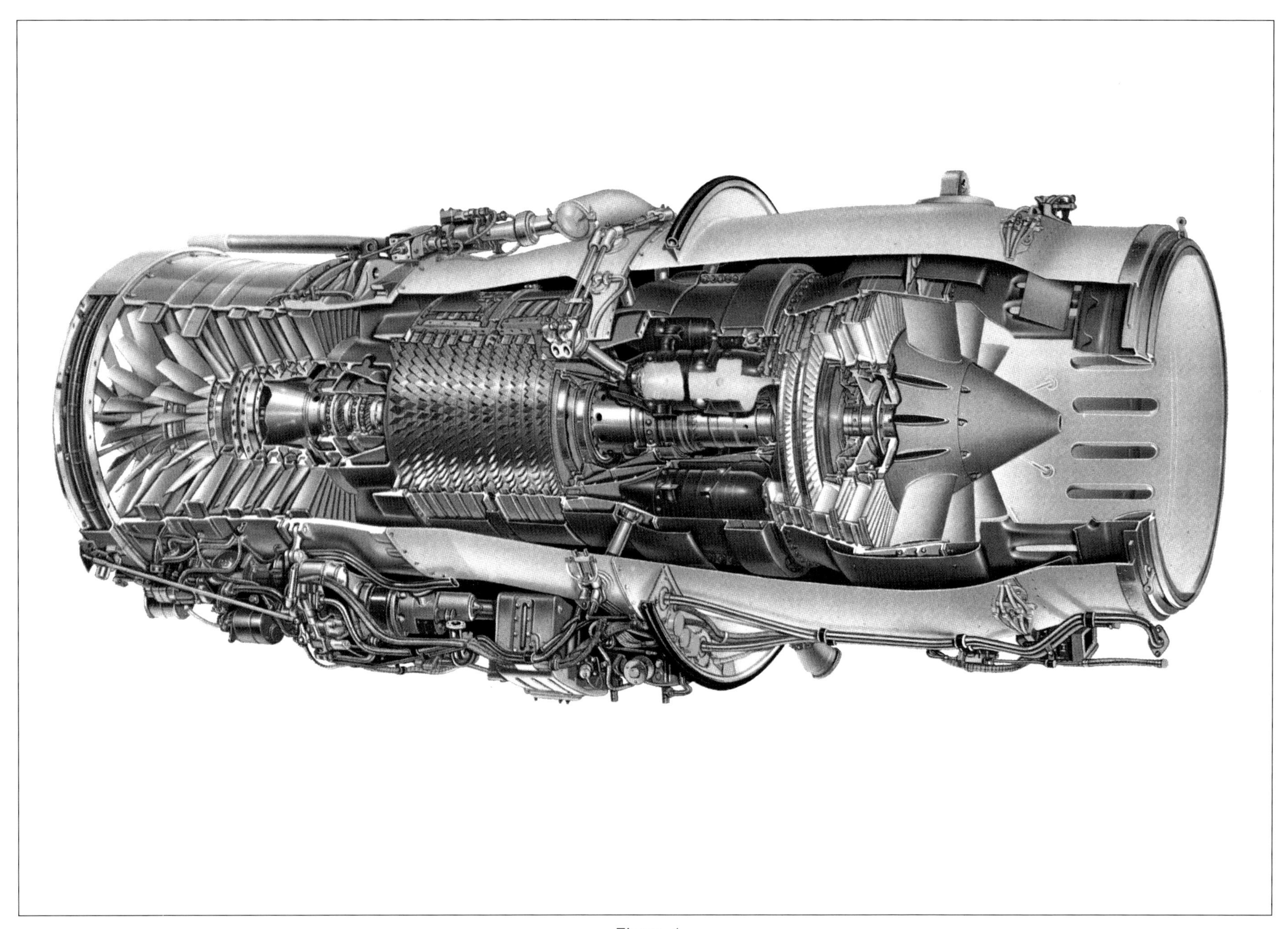

Figure 1
Rolls-Royce Spey Turbofan

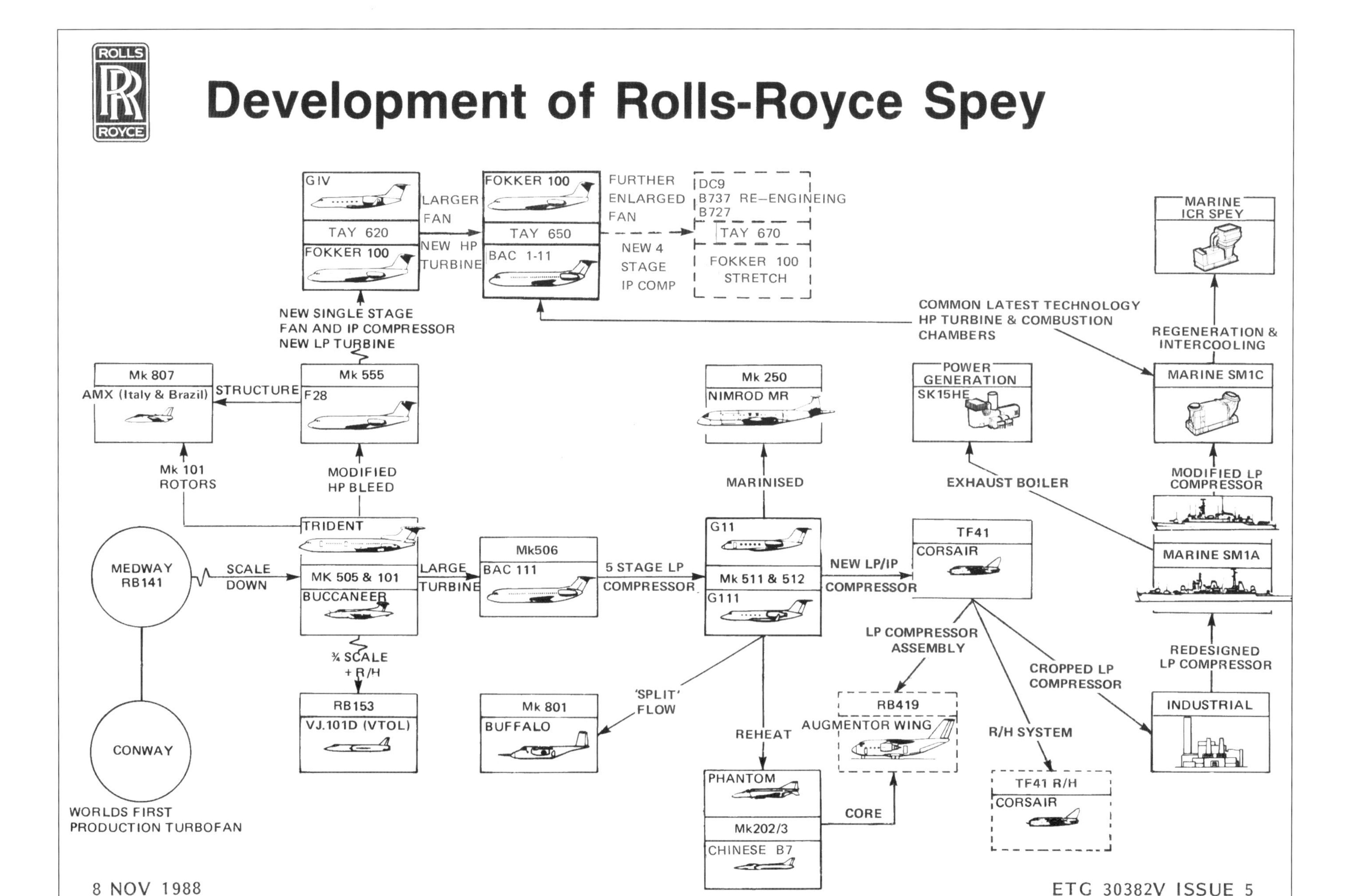

Figure 2
Development of Rolls-Royce Spey

INTRODUCTION

The Spey was the second Rolls-Royce bypass jet engine for commercial transport use entering service in 1964, six years after the Conway and eight years before the RB211. It can be argued that it was the first bypass engine in the world specifically designed for commercial use, though of course it later fulfilled a whole variety of military roles. In all its guises it has become the most versatile of all the Rolls-Royce family of engines, covering a variety of civil and military aircraft involving reheat, flap-blown carrier-born, maritime patrol, research and executive aircraft plus marine and static power roles - it even holds the world land speed record!

A very simple history

The initial requirement was for a short range commercial transport engine cruising at 0.875 Mach number. The key launching aircraft was to be the three-engined de Havilland DH121 for British European Airways (BEA). The Conway was too large so, in 1957, the RB141 (later named Medway) was designed, initially at 12,000 lb takeoff thrust, with some increase in bypass ratio and flame temperature, relative to the Conway, for operating cost and noise reasons. Development testing started in late 1959.

Then, in 1959 before the RB141's first run, on the basis of revised traffic figures, BEA requested a scaled-down Trident, so the RB163-1 (later named Spey) was designed at 9850 lb. Again the bypass ratio was increased and other changes made so that the engine was fine-tuned to BEA's new economic requirements. Development testing of the RB163 started in late 1960, and RB141 testing was reduced to support of the RB163. Getting the performance of the RB163 right proved to be a major task.

In 1959/60, Rolls-Royce collaborated with the Allison Division of General Motors on the definition of AR963 engines and on reheat development testing. Meanwhile, Boeing - learning something from the Trident - designed their own three-engined 727. The AR963-6, a 20% scaled-up RB163 was submitted to Boeing at 12,170 lb, who selected it and so did United Airlines but, thanks to Eddie Rickenbacker of Eastern Airlines, the 14,000 lb Pratt & Whitney (P&W) JT8D eventually won the competition. In addition, a reheated version of the AR963 was entered into the competition for an engine to power the swing-wing TFX (later the F111 aircraft); P&W won that competition with a reheated version of the JT8D (TF30).

Following the initial installation, the Spey was rapidly uprated and derated and found other civil and military installations in enlarged versions of the Trident, BAC One-Eleven, Buccaneer, F28, Gulfstream, Nimrod, Phantom, Corsair, AMX and a Chinese application. Rolls-Royce Canada had responsibility for another Spey derivative in the Buffalo, a research augmentor wing aircraft. These upratings needed significant modifications, including major redesign of the LP compressor and the introduction of water injection and reheat. It finished up at 12,500 lb civil, 15,000 lb military, and 20,500 lb maximum reheat for the Phantom. The Spey was mostly involved with traditional Rolls-Royce markets and aircraft constructors. However, in the BAC One-Eleven it penetrated the important North American civil market, and in the military market with the Rolls-Royce/Allison TF41, which replaced P&W's troubled TF30 in the Corsair for the US Navy and Air Force. The reheated RB168-25R was fitted to the McDonnell Phantom for the Royal Navy and Air Force and was also supplied to the Chinese Government.

More severe noise rules limited the life of civil Speys. In 1972 when the RB211 was under development, design work was undertaken on a re-fanned Spey, the quieter more economical -67. There was no market for it then, but ten years later the higher bypass ratio Tay (with a Spey core) was successfully launched. East Kilbride had engineering responsibility for the Tay, which had a future for new installations in regional and executive aircraft and re-engining existing types.

The Spey had also been adapted for industrial and marine applications at Ansty and achieved considerable sales, particularly in both British and foreign navies.

As a final pinnacle in a long and illustrious career, two ex-Phantom reheated Speys enabled Richard Noble to gain the world's land speed record and, in the process, break the sound barrier on land.

Engine types

All Spey aero engines have two shafts, the bypass air mixing with the hot exhaust in a common jet pipe (except the Buffalo). The industrial and marine versions have the bypass system removed, and a free power turbine added.

	Compressor Stages	Turbine Stages
RB141-11	5 LP + 11 HP	2 HP + 2 LP
RB163-1, -2, -2W, RB183, RB168-1	4 LP + 12 HP	2 HP + 2 LP
RB163-25, RB168-20, RB168-25R	5 LP + 12 HP	2 HP + 2 LP
TF41	3 LP + 2 IP + 11 HP	2 HP + 2 LP
Tay	1 LP + 3 IP + 12 HP	2 HP + 3IP
Industrial & Marine	5 LP + 11 HP	2 HP + 2 LP + 2 power turbine stages

The Tay and some industrial and marine types are not yet history, so the major part of the book covers the years 1958 to 1970. Because of the many variants of the Spey and size limitations, it can only deal with engines which reached the hardware stage – ie service, development or demonstrations only – no paper engines. The RB153-61, which was a 78% linear scale of the Spey RB168-1 fitted with reheat and swivelling nozzles for vertical takeoff, did reach the development stage but did not go into production and thus is not mentioned further in this book; it was a very interesting and complex engine and could be the subject of a separate book. Of course, many configuration possibilities were investigated:-

- Variable mixer and final nozzle areas for hot day takeoff boost, particularly for boundary-layer bleed engines.
- Reheat of bypass air only.
- Front fan, aft fan, front and aft fans at the same time.
- Double bypass.
- Inter-cooled regenerative marine engine (This study led to the WR21 marine engine).

Nomenclature

In the book, engine types are referred to by their ratings, mark numbers or even aircraft installation. The following list of aero engines, which went into service, may help to avoid confusion. It is by no means the complete list, but it covers the major engine types and aircraft installations. In total there were 94 mark numbers/ratings symbols, which also covered particular airline ratings and installation features.

RB163-1	Mk 505	Trident 1
RB162-2, -2W	Mk 506, 506 AW	BAC One-Eleven 200
RB163-25	Mk 510	BAC One-Eleven 3/400, Gulfstream II and III, Trident 1E
	Mk 511	Trident 2E and 3B (3 Speys and 1 RB162)
	Mk 512	BAC One-Eleven 475/500
RB183-2 (Spey Jr)	Mk 555	Fokker F28
RB168-1	Mk 101	Buccaneer
	Mk 807	AMX
RB168-20	Mk 250/1	Nimrod
RB168-25R	Mk 201/2/3/4	Phantom, Chinese application
RB168-62	TF41 A1/2	LTV A7 (Corsair)
(-25 type)	Mk 801 SF	Buffalo
Tay	Mk 611	Gulfstream IV
	Mk 620	Fokker 100
	Mk 650	Fokker 100, BAe One-Eleven re-engined, Boeing 727 re-engined

The above rating numbers are the major listings. There were many variants of each major number – in total about 94. A section dealing with the many different mark numbers appears in the Appendix.

Industrial and marine versions:

Marine Spey	SM1A	Various naval vessels for UK, Japanese and Netherlands navies
	SM1C	Uprating and redesign of SM1A
	SM1CR (study only)	Inter-cooled regenerative engine for marine use – from which WR21 derived
Industrial Spey		Gas generation, oil pumping, power generation and process industry

Chapter one: DESIGN

The Spey was the eventual outcome of studies in the period 1957-60 to find the optimum engine to power short to medium range subsonic civil transport aircraft; many sizes and variations of the basic cycle were looked at. The basic requirement was for an engine to power an aircraft which had been specified by BEA. It was required to fly from London to Paris at Mach 0.875. Several British aircraft manufacturers were involved in the competition, most proposing four-engined aircraft, but de Havilland – as a result of a detailed evaluation involving Rolls-Royce – proposed a three-engined aircraft. It was deemed to have lower operating costs than the four-engined aircraft.

The engines offered for these competing aircraft designs were the RB140 and the RB141, the former rated at 8,000 lb thrust for the four-engined aircraft and the latter rated at a nominal 12,000 lb thrust for the three-engined aircraft. The winner of the competition was de Havilland with its DH121 and the engine was to be the RB141; this engine was also considered for a growth version of the Caravelle. After some time, when the programme was well underway, BEA decided that the aircraft was too large and, early in 1959, called for a scale-down. The smaller engine that the aircraft now required was the RB163, later named the Spey. Although its primary installation was the DH121, eventually called the Trident, many other applications were to follow, some long range, some supersonic and some military.

Following the success of the earlier Conway, which was the first bypass engine in the world to go into commercial service (in the Boeing 707), there was no doubt that the Spey would be a two-shaft bypass engine, but optimised for smaller two- or three-engined aircraft of shorter range, say 200 to 1500 miles, and speeds up to Mn 0.875. The important requirements were:-

1. Low operating costs – a mixture of low installed specific fuel consumption (SFC), weight, cost and good reliability. The latter three were more important on short range.
2. Low takeoff and landing noise, especially from short runways, even though the Spey pre-dated the era of noise certification.

Design features described in this chapter mainly refer to the initial Spey RB163-1 in the DH121 Trident. Of course, other installations and upratings had different cycles, aerodynamics, materials and construction. These will be described in the appropriate chapters later in the book. Many of the Spey mechanical features were fed into the later development of the Conway, the RCo42.

Choice of cycle

The cycle parameters, which needed to be optimised, were bypass ratio (BPR), overall compression ratio (OPR) and turbine entry temperature (TET), together with a decision about mixing the bypass air with the turbine exhaust.

The optimum values of the three parameters are dependent on each other. For instance the optimum BPR is higher at higher TET, but higher OPR makes a higher TET more difficult because of the hotter blade cooling air. In practice, the OPR and TET chose themselves at values somewhat above the Conway, taking advantage of improvements in materials and aerodynamic and blade cooling technology.

The takeoff OPR of 16.8 was some increase over the Conway to give lower SFC. It was limited by turbine blade cooling requirements and avoidance of more expensive materials at the back of the high-pressure (HP) compressor.

The takeoff TET of 1319°K was similar to the Conway. Improved turbine cooling and materials were used to compensate for more severe cyclic duty, the higher percentage of time at takeoff per flight, and higher climb and cruise ratings relative to the Conway.

The original Conways were installed within the wing, a factor possibly influencing engine diameter and its original choice of 0.6 BPR. The podded Spey had no such size limitation. The influence of SFC on operating cost was less because of the shorter range, and also oil prices had not then rocketed. The curve of operating costs vs BPR was calculated to be virtually flat between 0.7 and 1.2. Other considerations chose 0.7 for the RB141 and later 1.0 for the Spey. The design of the RB141 had shown that the weight penalty for going to a higher bypass ratio was not quite as severe as anticipated and also noise was getting more important so it was worth the cost of increasing the Spey BPR to 1.0 to reduce jet velocity and hence takeoff noise. There was no point in going higher because the noise emanating from the compressor and elsewhere predominated at approach. (For a more detailed discussion of the choice of bypass ratio see the Section 1 in Afterthoughts).

Depending on the cycle, there is a potential 4 to 6% of thrust to be gained by complete mixing to a uniform temperature of the two streams in the jet pipe; for a practical mixer this figure is halved due to pressure losses. To realise this gain and to enable both streams to be reversed and to reduce the hot velocity and hence noise, the Spey was designed with a mixer.

The Conway came about six years after the Avon, and was a large step in BPR, OPR, TET and the number of shafts. On the other hand, the Spey came six years after the Conway and had much more modest increases in these parameters. For those who had to make the decision, the Spey was considered the optimum to meet the requirements with the technology then available. The RB211, which was still eight years away, was to be a very large step, particularly with the BPR going up from 1 to 4.5. By then many things had changed; the prospect of more severe noise regulations, advanced compressor aerodynamics and turbine blade cooling, weight reduction and lower pod drag, to say nothing of stronger competition. Details of the cycles for Rolls-Royce engines and competitors to the Spey are given in Table 1 in the Appendix.

Aerodynamics and layout

In order to achieve the mixing gain, the hot and cold streams have to be at roughly equal total pressures, corresponding to nearly equal low Mach numbers and, therefore, minimal losses. On the RB141 this fixed the low pressure (LP) ratio at takeoff at 2.6, leaving 6.4 for the HP. As compressors over about 5.5 pressure ratio require variables for low speed operation, the HP compressor had a variable inlet guide vane and handling bleed. Some thought was given to an LP/IP (intermediate pressure) arrangement, with some of the HP compression on the LP shaft. On early Conways this had been avoided, due to its lower BPR and because with separate jets a Pcold/Phot of 1.35 was acceptable. Later Conways had mixed exhausts and this resulted in very high mach numbers in the cold side of the mixer. The LP/IP principle was not adopted on the RB141 or RB163 because of worries about IP surge due to changing incidences on the splitter, and we had learnt a lot about variables on the Avon where they had saved the day. The Conway Co42 and 43 with LP/IP compressors and mixers came later. Our two P&W rivals, JT8D and TF30, and then the Spey TF41, all had LP/IP configurations.

The aerodynamic design of both compressors was based closely on the successful Conway, eg, 62% reaction, radial equilibrium, C4 blade sections, VIGVs, etc; some progress had been made in designing for slightly increased pressure ratio per stage and this saved the equivalent of 1½ stages.

Two potential improvements were not yet proven; a transonic LP compressor without IGVs, which was designed on the TF41, and low aspect ratio blading (wider chord) which came in later on the Adour – though, in the final event, the compressors for that engine were designed by Turbomeca, Rolls-Royce's partner.

On the Avon a satisfactory surge line was ultimately achieved with variable IGVs and bleed from two mid stage stations overboard where, in performance terms, it was written off. The Spey was to follow suit, but with a better bleed arrangement where air was extracted uniformly round the circumference instead of two radial positions, and then exhausted neatly into the bypass duct where some thrust was recovered.

An alternative to bleed, and favoured by some, was to have several rows of variable stators. The principle (an invention of Geoff Wilde, Head of Compressor Department, in 1947) had been tested on an Avon rig, many years before the General Electric (GE) J79, which had seven variable stages, and later on the 12-stage

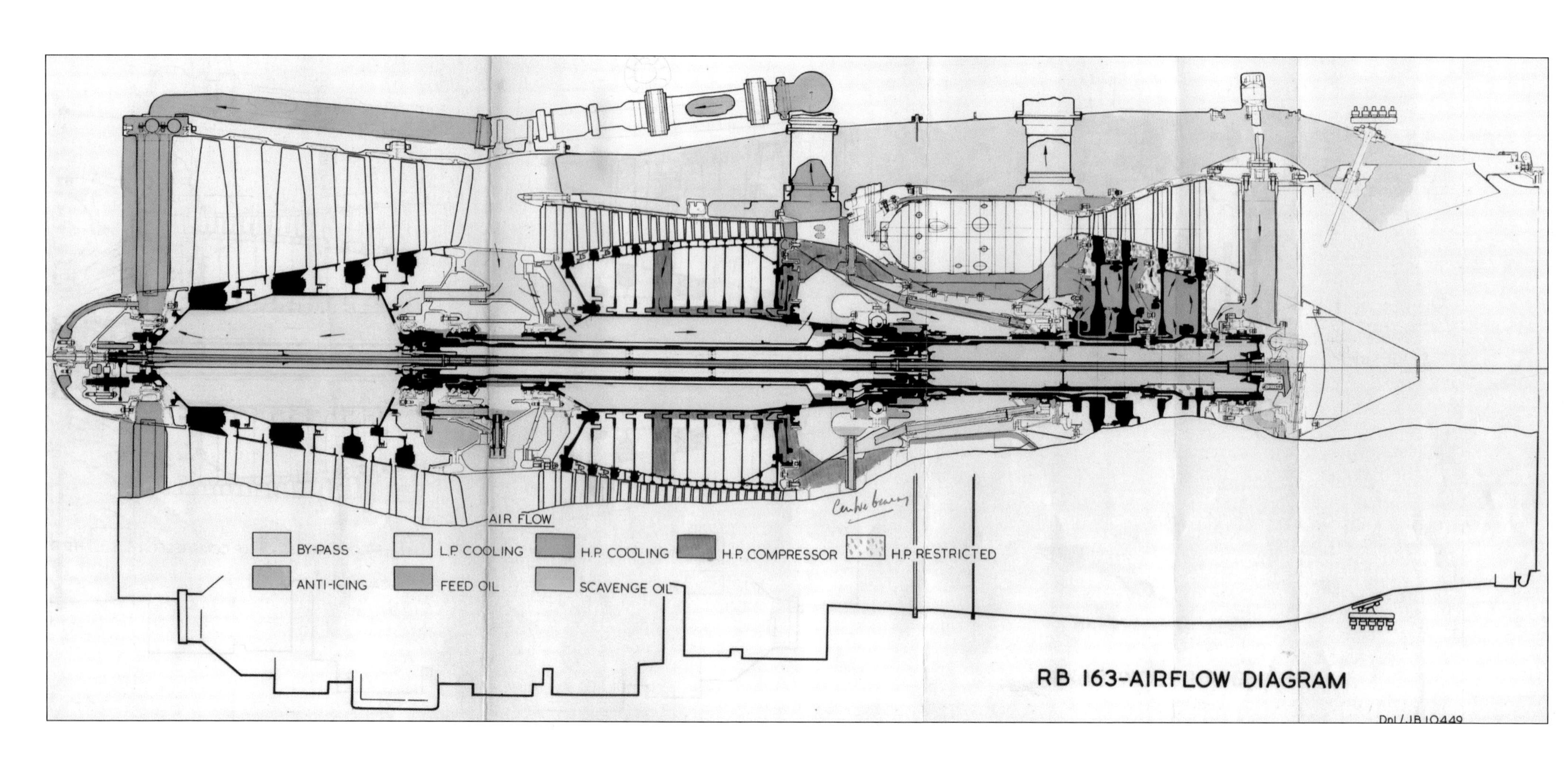

Figure 3
RB163 Airflow Diagram

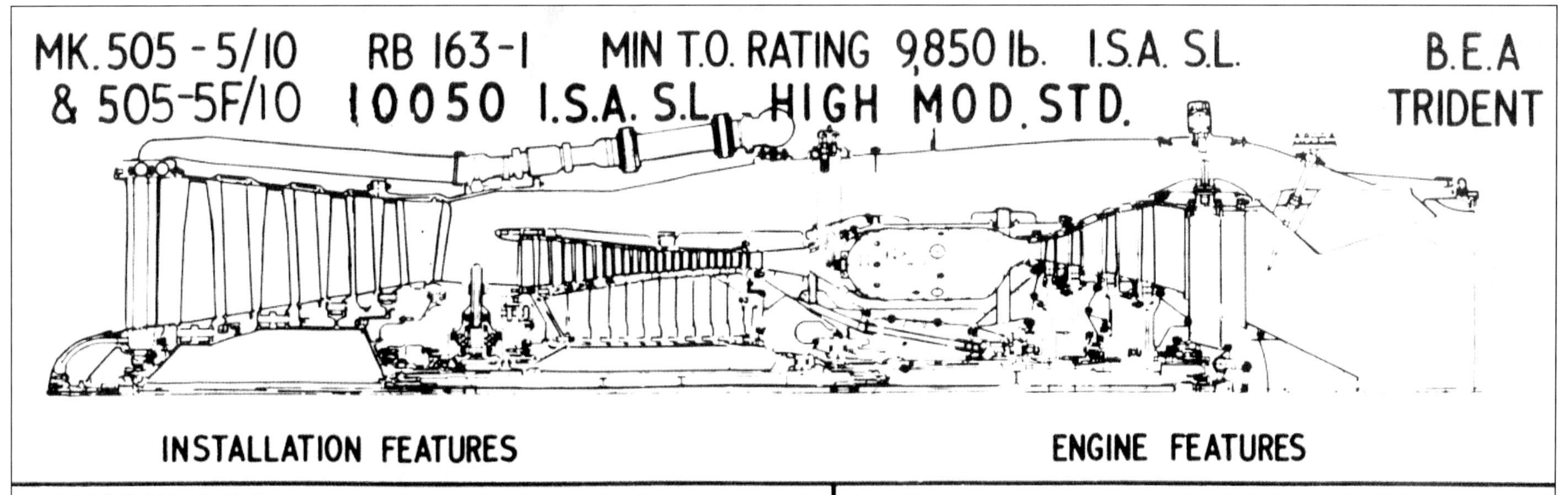

- HOBSON C.S.D.
- ROTAX STARTER
- SPERRY HYDRAULIC PUMP V CLAMP FLANGE
- FUEL PRESSURE TRANSMITTER
- ELLIOT FLOWMETER
- ROTAX IGNITION SYSTEM 2×12 JOULE SYSTEM
- TERMINAL HARNESS

- 4 STAGE L.P. COMPRESSOR
- 12 STAGE H.P. COMPRESSOR
- 2 STAGE L.P. TURBINE
- 2 STAGE H.P. TURBINE

Figure 4
General Layout of Original Spey

Spey rig itself where IGVs and four rows of stators were tested at fixed positive staggers. The results were promising, but it was felt that the additional complication plus some mechanical problems did not warrant the benefits. Further aerodynamic optimisation work would have been needed and there was insufficient time to do this and design the complex operating mechanism.

Imaginative thoughts were given to using the bleed to further benefit performance rather than wasting it. The theory was that bleed should be recirculated to the front of the compressor and injected at the tip to unstall the front stages at low speed. A test on an RA14 engine gave poorer acceleration. Then a rig test was run on an RA29, an annular nozzle was sized to shoot the air in at the tip, but even with the velocity over the outer 1/10 of annulus double that over the inner 9/10, there was no effect on surge line or anything else (other than that which would have occurred with a uniform increase of inlet temperature). A theory that it should have been injected at the root was quickly invented. Thankfully, this design nightmare was not put to the test.

The LP/HP compressor split led to HP and LP turbines of two stages each. The aerodynamic design was based on a 1957 correlation by Stan Smith of all turbine efficiencies to date. (The Specific work – Δ H/U^2 versus Specific Flow – Va/U chart is still in use today though much modified). The design parameters were work output, blade speed and axial gas velocity. The RB141 was designed along the ridge of the efficiency contours at the best end. Blade numbers were chosen from a correlation of space/chord versus deflection angle. The RB141 efficiencies came out better than expected, but the RB163 was to be a different kettle of fish.

The bypass duct was designed with low entry Mach number levels to keep the pressure loss down. For installations with long jet pipes, an annular mixer obtained adequate temperature mixing, with a substantial part of the thrust gain without large pressure loss. On the RB141 the short pipe necessitated a forced mixer with its attendant pressure loss. The design was a 20-chute injection mixer with the bypass air forced in at a steep angle.

Figure 5
Front View of Spey LP Compressor, showing inlet guide vanes offset at root to minimise thermal stresses

It should be noted that a two-shaft mixed bypass engine has four degrees of freedom. Desired matching of compressor speeds and pressure ratios is normally achieved by adjusting HP and LP turbine capacities, final nozzle and mixer areas. Some of these procedures were to cause considerable troubles during RB141 and RB163 development.

Mechanical design

Following the reduction in the size of the DH121, as described above, the Spey replaced the RB141. Although basically a scale with increased bypass ratio, there were mechanical design improvements aimed at reducing the weight and cost. However, the basic concept of the design of both engines was taken from the earlier Conways (RCo12), which had the longest overhaul life of any engine anywhere. Principal objectives of the mechanical design were to give reliable operation, long overhaul life and low cost maintenance, and also to maintain good efficiency levels and surge margins by minimising steady running and transient clearances. Among the following selection are many features novel to the Spey, and copied on later Conways.

Stress concentrations were avoided, particularly in sheet metal components. The stainless steel front bearing housing was an example. The IGVs were non-radial, so that when heated by anti-icing air they would tend to rotate the hub rather than bend or punch holes in the casing. The end attachments of the vanes were away from the extreme ends, and butt-welded instead of overlapping joints. Oil feed and scavenge of the front bearing housing was through the centre of the shaft, to avoid non-symmetry of the vanes. Another novel feature was the anti-icing hot air manifold with holes in the pipe, clamped directly on to saddles on the outer ends of the vanes: Don McLean says it was copied from his Ford Zodiac exhaust manifold.

Measures were taken to reduce foreign object damage, particularly birds. As a result of Avon and Conway experience, a large gap of 2¼" between IGV and LP1 rotor was designed and, unlike early Conways, did not have an overhung LP1 disc in front of the bearing. The LP compressor rotor was a simple light drum construction in two parts bolted together at the last stage. As it also minimised the out of balance, it was chosen on the 163 in preference to the shaft and discs on the 141. The longest compressor stator blades in LP and HP stages 1-3 were shrouded for strength and to minimise leakage. Early engines had the remaining HP stators, two stages at a time, dovetailed into steel half rings. They suffered from fatigue failures, frettage due to circumferential movements on acceleration and deflections on surge. The design was completely changed to single row T-slot fixings with semi-circular tightening straps. This also gave a benefit in surge pressure. Both compressor casings were split; for lightness the LP was in magnesium steel, protected from hot air by the stator rings and platforms.

To begin with the 10 flame tubes had integral discharge nozzles, but half way through development they were made detachable for ease of repair. On later models combustion casings were split to facilitate inspection of the flame tubes and HP1 NGVs and turbine blades without stripping the engine.

Forged HP1 blades were in Nimonic 105, with cooling air pre-swirled and ejected through 5 holes at the tip; HP2 blades were uncooled in N 105. The blades had extended shanks and fir tree roots to keep the disc rims cool. HP1 and 2 NGVs were cooled.

Engine vibration can be the cause of failures of accessories, cracking of structures and frettage of mating surfaces. To avoid this, the two roller bearings on the turbine shafts were carried from the engine structure by a spring-mounted member, comprising a ring of axial rods rather like a squirrel cage. This reduced the transmission to the engine carcase of vibration resulting from residual out of balance of the rotating parts. For the same reason, later Speys had a squeeze film bearing at the front of the LP shaft.

Oil was fed from the rear of the engine through a long pipe in the centre of the LP shaft, and then to all the main bearings. These bearings were blanketed with a small inflow of LP cooling air, which kept them cool, and also the oil, which reduced oil consumption. The load on the thrust bearings was adjusted by seal diameters to be a minimum, but consistent in direction. A small amount of HP

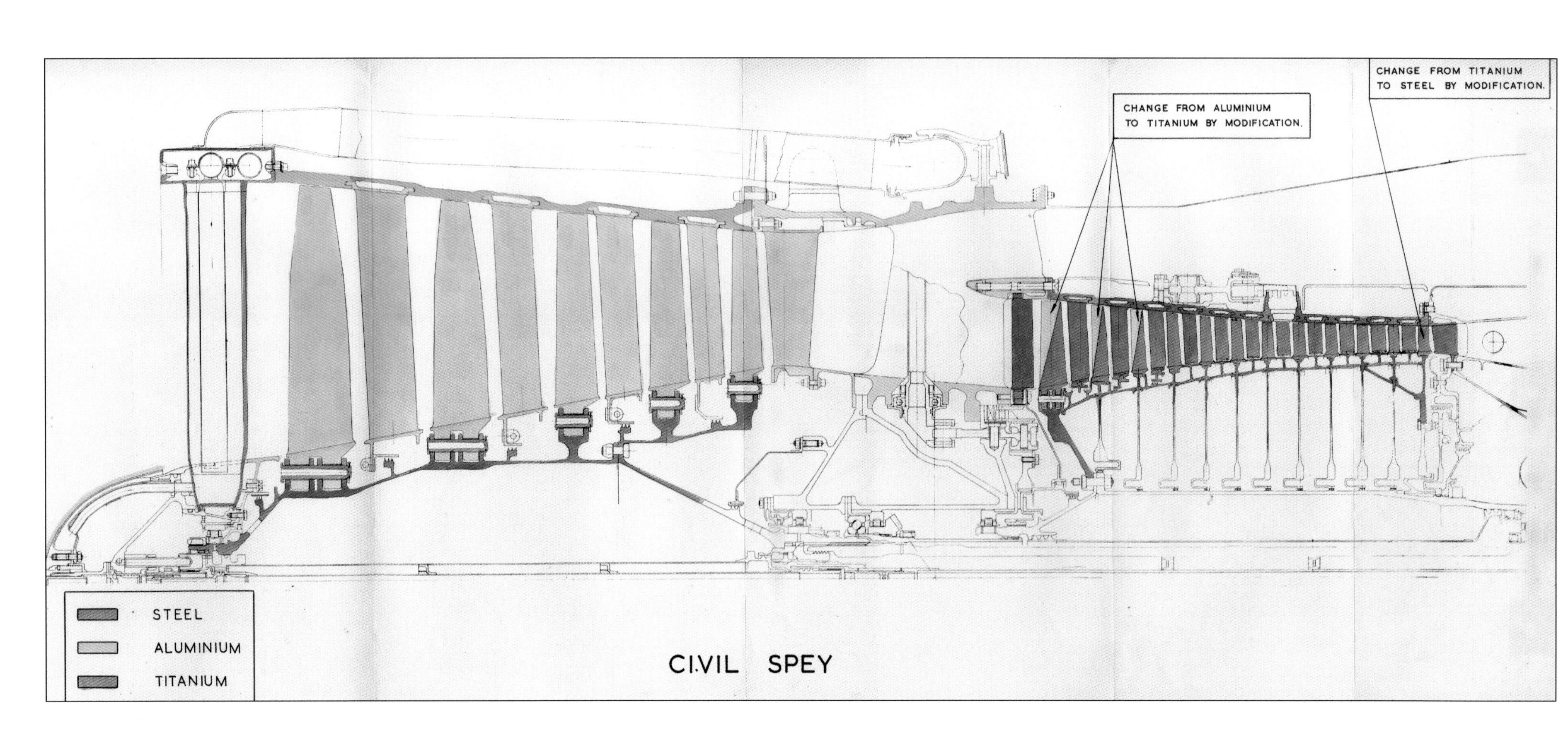

Figure 6
Mechanical Construction of LP & HP Compressors

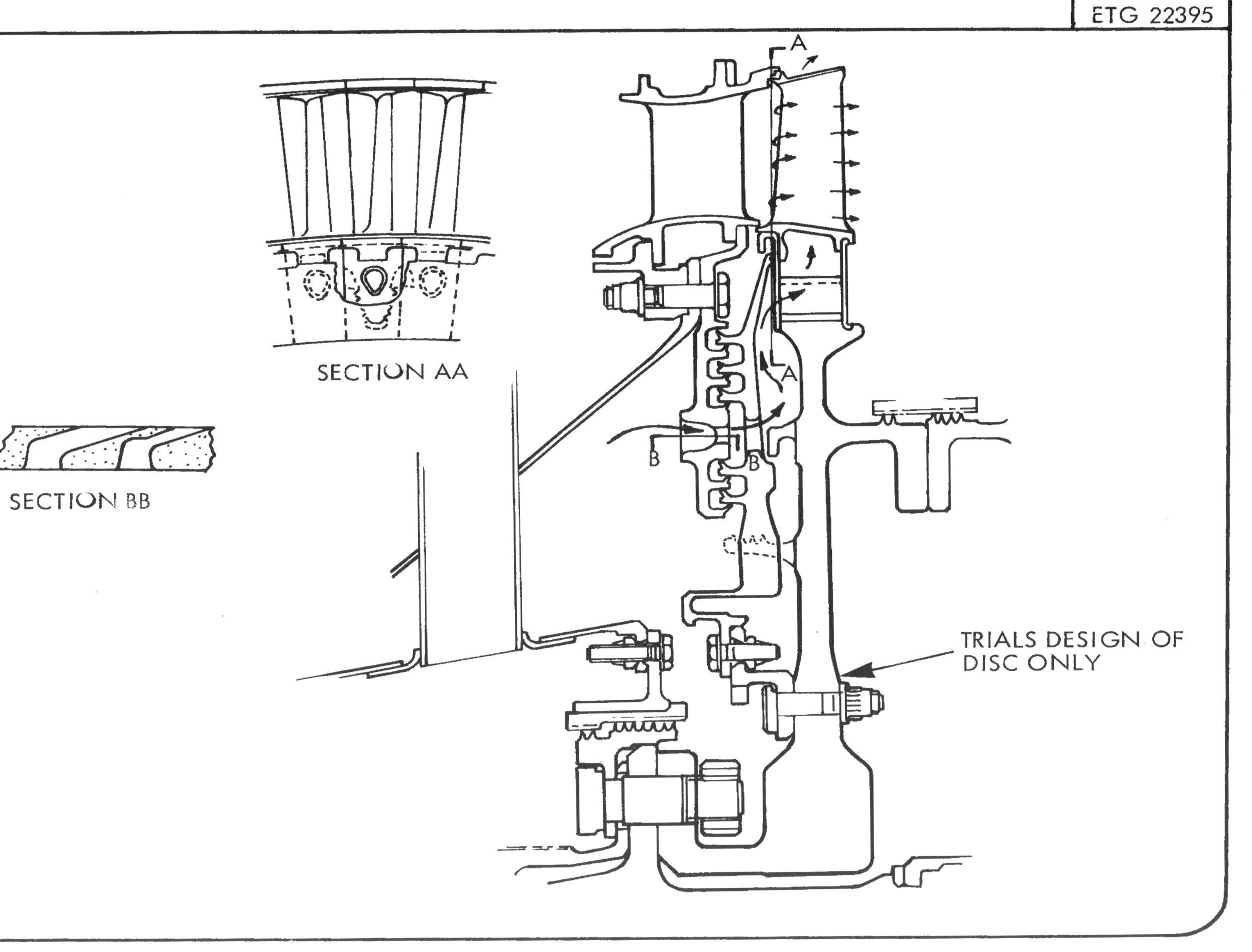

Figure 7
Spey - HP Cooling Air Feed to HP1 Turbine Blade

delivery air was used for sealing the turbine, preventing hot gases from flowing inwards on to the discs. Attention was paid to sealing the spaces between LP and HP air and HP5 air was used to cool the rear compressor disc roots.

Both LP and HP air was supplied to the aircraft for cabin air and other uses. This air was free from oil contamination as a result of the air-blown bearing chamber seals, and an outer seal on the HP compressor front bearing, which was vented overboard.

To guard against the potentially catastrophic results of an LP shaft failure, however unlikely, it was necessary to have a fuel trip mechanism to prevent the now unloaded turbine from accelerating up to disc bursting speed and subsequent non-containment; such a failure had occurred on the Conway. The device had to be totally passive to prevent inadvertent operation, and only had a minute fraction of a second to work. The long central oil tube (which was assumed to remain in one piece after a shaft failure) was splined to the shaft at the front, and to a nut with a quick-start thread at the rear. In the event of a shaft failure, within one excess revolution of the two ends of the oil tubes the nut was driven out and struck a lever. The lever was connected via a long cable (known as the 'bog chain') and pulleys to the fuel shut-off cock. The virtually instantaneous reduction in fuel was sufficient to slow down the turbine before it reached burst speed.

Most of the accessories were positioned with access from below in such a way that each could be removed or replaced separately without disturbing anything else. On the Trident Rolls-Royce had technical and supply responsibility for the complete power unit including nacelle, although the nacelle aerodynamics were determined in association with de Havilland. The thrust reversers had outlets designed to avoid throwing up debris into the intakes, and six lobe silencers fitted at the back of the pod engines.

The two side fuselage engines were mounted from two curved beams, attached to aircraft stub wings in the plane of the compressor intermediate casing and LP turbine exhaust casing. They supported the engine rigidly, while swinging links and ball joints from the turbine casing permitted it to expand and contract as necessary. An axial strut transmitted the thrust to the aircraft structure.

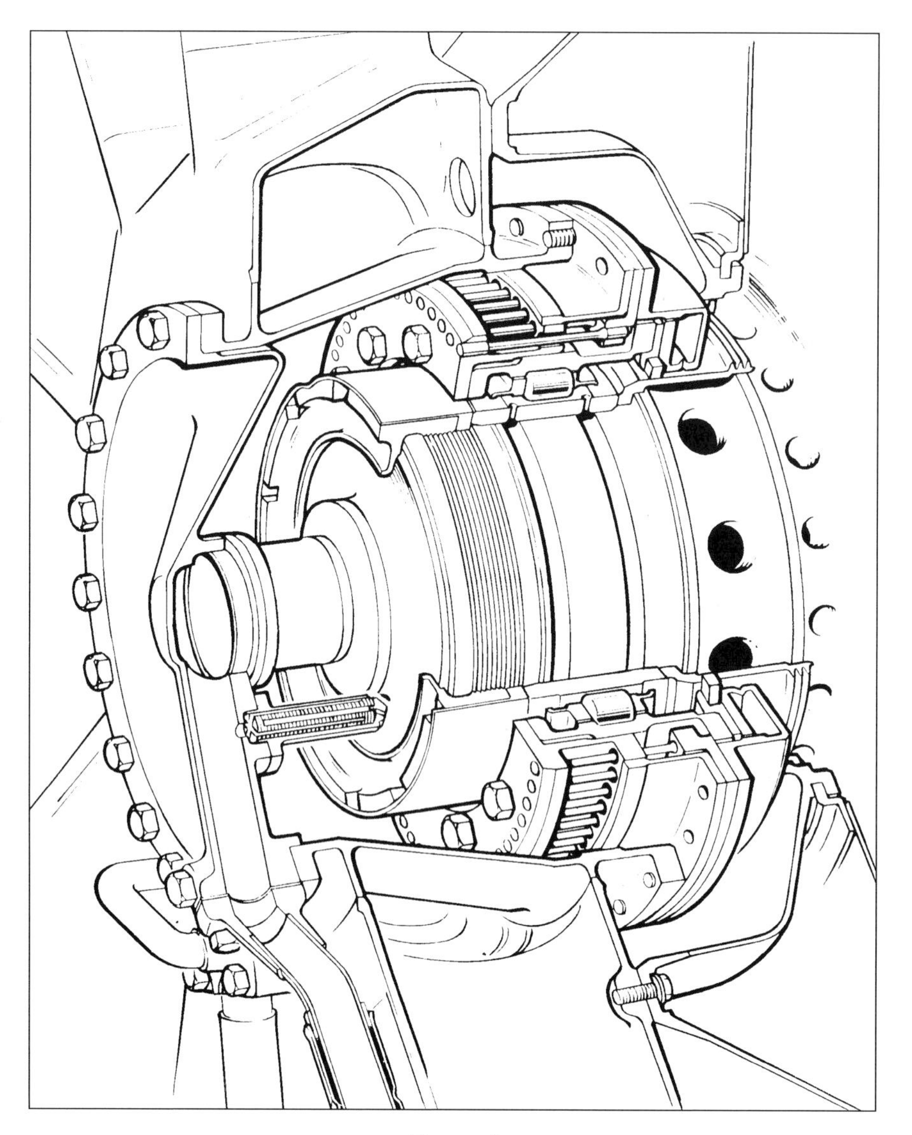

Figure 8
"Squirrel Cage" Mounting of Roller Bearing

Figure 9

Trident I at Hatfield

Figure 10
External view of RB163

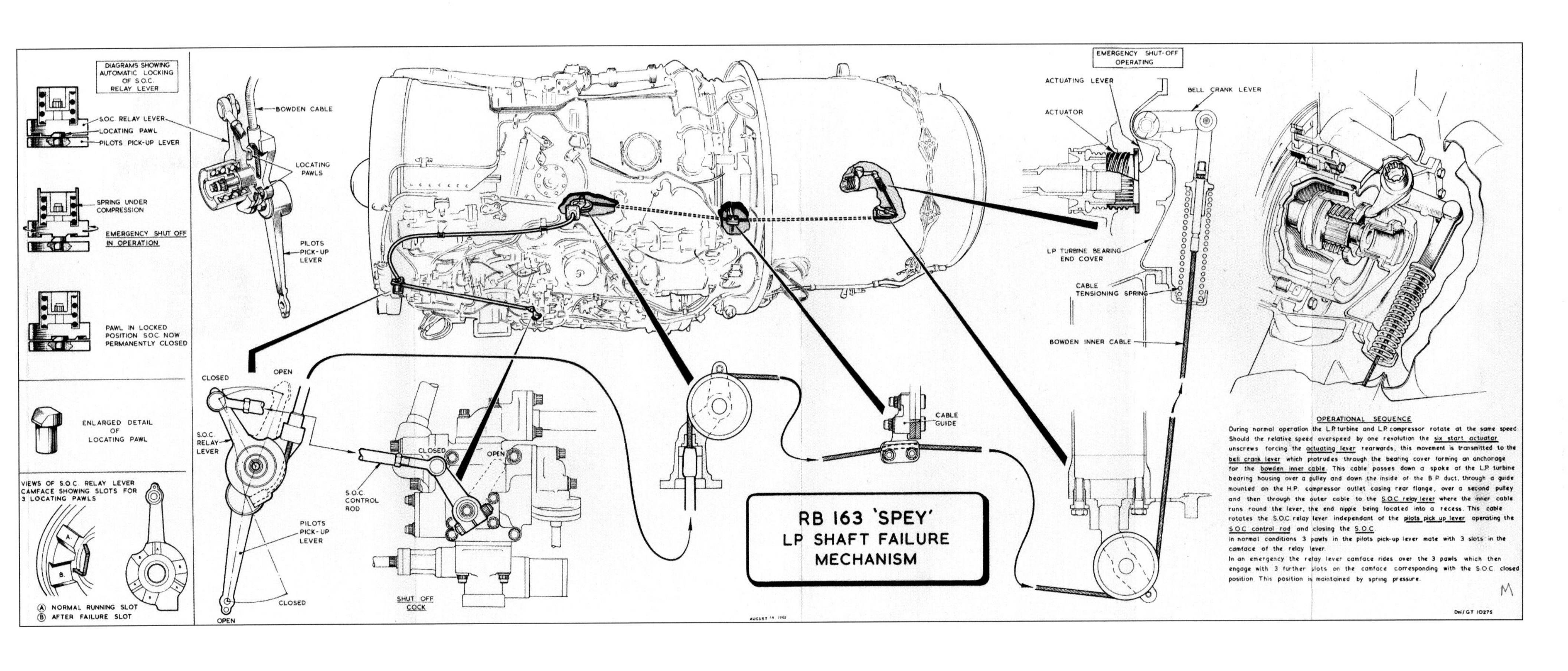

Figure 11
RB163 'Spey' LP Shaft Failure Mechanism

Fuel control system

Introduction:

The Spey fuel system was the result of an appraisal in the 1950s of the needs of modern aircraft for reliability, an acceleration control to exploit compressor surge margins over a range of flight conditions, and accuracy of governors and limiters. The resultant fuel control system for the Spey broke new ground in three different ways, listed below. For this reason, a complete description of it and its method of operation are given in this section.

(a) Lucas not only took responsibility for the design and development of the fuel system, as they had on previous engines, but also for the funding.

(b) The control parameters and laws were very different from those used on previous engines.

(c) The basic design of the controller was radically changed in concept from previous systems which had employed metering valves controlled by 'hydraulic computers' involving small flow servo elements.

Five governing and limiting parameters were needed, some to limit operation within the type-tested conditions whilst ensuring the achievement of certificated thrusts, others for ease of control etc:-

- HP shaft rpm (NH) – The basic control parameter. It was desirable to make it isochronous, ie, a fixed throttle to give a fixed NH for most conditions.
- LP shaft rpm (NL) – To protect the LP system from overspeed and prevent excessive non-dimensional conditions at low inlet temperatures.
- Compressor Delivery Pressure (P3) – To limit the maximum pressure in the engine and flat rate the takeoff thrust at low altitudes on cold days, to preserve engine life.
- Turbine Gas Temperature (TGT) – Needed to limit TET(T4), particularly with air offtakes. Scatter in the T4/T6 relationship led to TGT(T6) trimming, and later to improved positioning of the 10 thermocouples.
- Acceleration and Deceleration Fuel Flow Controls – To prevent surge/flame-out during acceleration/deceleration.

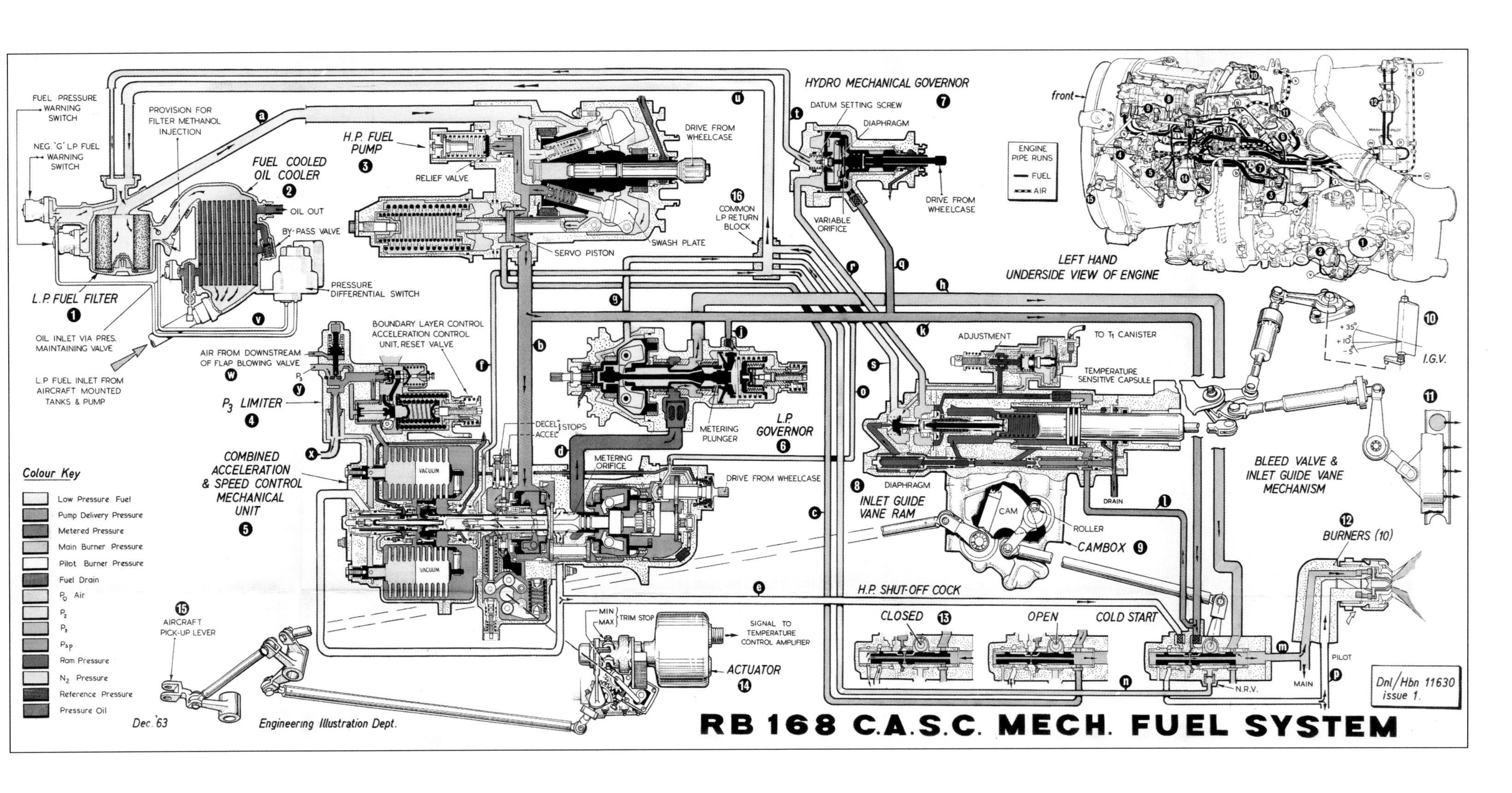

Figure 12
RB168 CASC Mechanical Fuel System

Control parameters and laws:

To achieve high acceleration rates without surging, the HP compressor required careful control of transient working lines throughout the flight envelope. Previous engines had fuel scheduled against compressor delivery pressure to keep constant fuel/air ratio to protect against over-temperature; not necessarily the best to match the surge line and thus obtain the best possible acceleration. Compressor working lines and surge lines can be expressed with acceptable accuracy in non-dimensional form, for example pressure ratio against non-dimensional airflow or speed, eg, $F/P_2\sqrt{T_2}$ and P_3/P_2 against $NH/\sqrt{T_2}$. From the measurements available, a schedule of F/P_2NH versus P_2/P_3 was selected to avoid the inaccuracy of transient temperature measurements – see Figure 13. Following throttle opening, fuel flow increased to a scheduled level below the surge line with allowance for air inlet pressure distortion, Reynolds number and other effects. As acceleration took place, increases in P_2, P_3 and NH permitted an increase in fuel flow according to the schedule, until governed speed was approached and fuel trimmed to give the steady state condition selected by the throttle. The deceleration schedule was a scale of the acceleration, but below the steady running line.

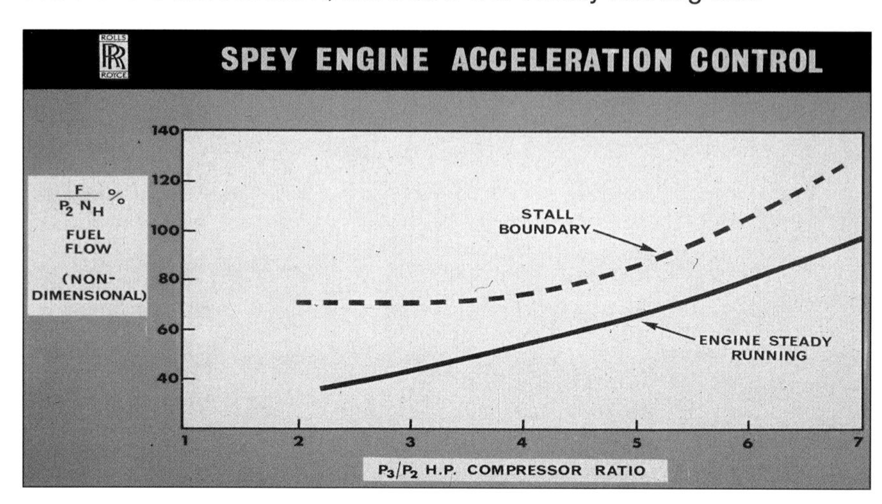

Figure 13

Mechanical design of the Constant All Speed Control (CASC):

Various ways of engineering the CASC principle into hardware had been considered by Lucas and Rolls-Royce and the system finally adopted was the CASCMECH, which was initially schemed by Rolls-Royce, principally involving Gordon May, Gordon Morley and Albert Jubb – the objective being a system free of hysteresis and insensitive to dirt and fuel specific gravity.

The initial Lucas had featured the usual half ball valve servo systems using the new control parameters, but Rolls-Royce, having experienced in-service problems with dirty fuel and small orifices, insisted that their design – using full flow metering and spinning valves to reduce friction – be used. It was thought that mechanical problems, such as wear on the rotating parts, would be easier to overcome than the problems of hysteresis and dirt sensitivity.

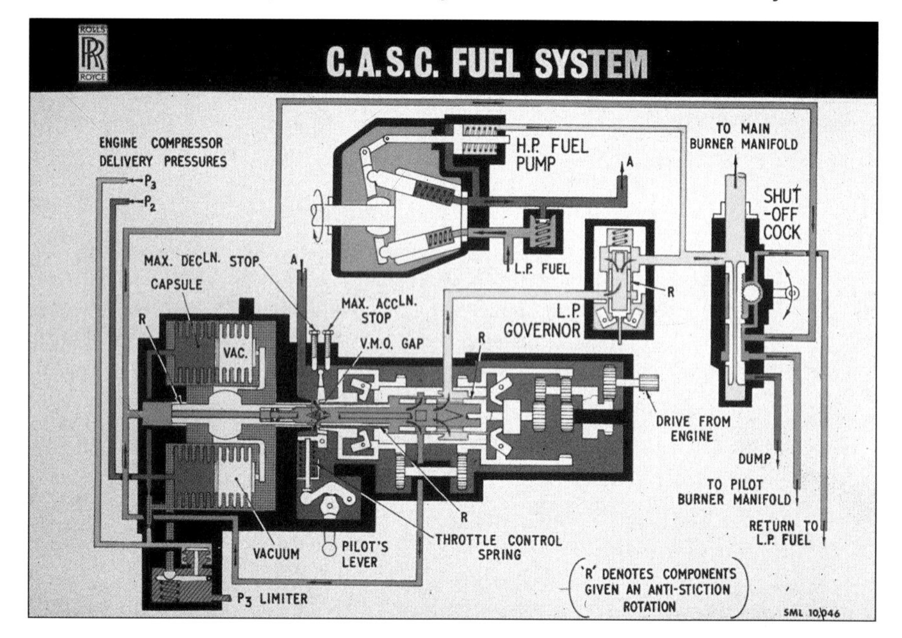

Figure 14

Figure 14 provides a diagram of the system and illustrates the co-axial disposition of the rotating valves.

The fuel flow to the engine provided by the fuel pump was regulated according to the pressure difference applied across the pump servo piston to vary the pump stroke; this pressure difference was arranged to be the combined pressure drop across two variable orifices in series in the engine fuel flow within the CASC.

The size of the first of these triangular orifices, known as the variable metering orifice (VMO), was controlled by a combination of two axial movements, that of the rotating valve and that of the rotating sleeve within which the valve slides and rotates. The position of the valve was controlled by the length of a capsule assembly, part of which was evacuated and subject to pressures which are a function of P_3/P_2 derived from the flowing air orifice system. Figure 15 shows the mechanical arrangement of the VMO.

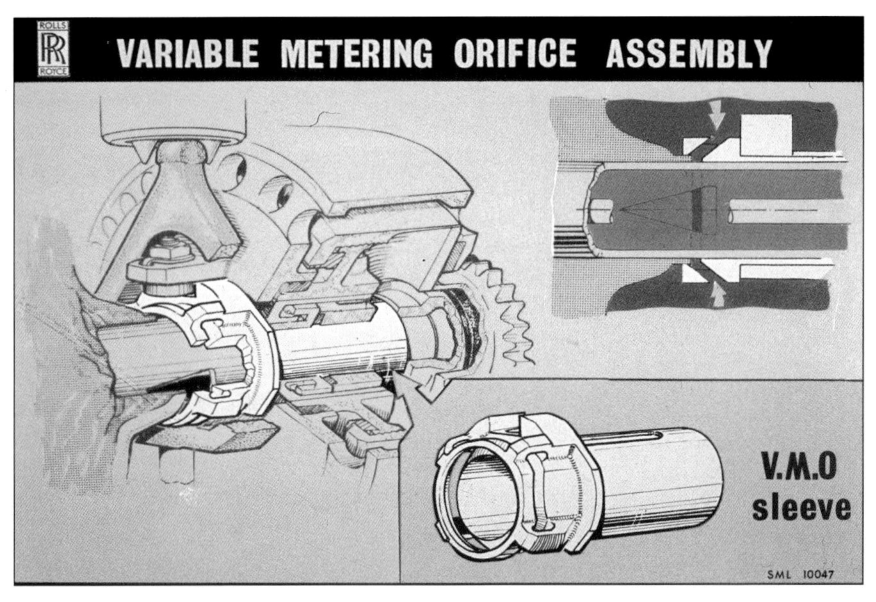

Figure 15

The position of the sleeve was controlled by a flyweight governor to control the HP shaft speed (NH) according to the position of the pilot's throttle lever but subject to the travel in either direction being limited to stops which provide the acceleration and deceleration calibrations, which are adjustable.

The second variable orifice was merely the means whereby the pressure drop across the VMO was controlled to be proportional to the square of the HP shaft speed. This was done by the pressure drop flyweight governor which, as illustrated, moves the valve so as to achieve a balance between the flyweight force and the load on the piston subjected to the VMO pressure difference.

In the acceleration and deceleration modes, the flowing air orifice capsule system produces a VMO movement proportional to $(P_3f(P_3/P_2) - kP_2)$ whilst the flyweights provide a pressure drop proportional to NH^2.

Combining these effects results in the desired characteristics viz:- $F/P_2NH = f(P_3/P_2) - \text{const.}$

The P_3 limiter, LP governor and SOV are not shown on Figure 15 nor is the electronic TGT limiter, which derives its safety inspired limited authority capability through operating on the NH governor maximum setting via an actuator in the linkage. Other control accessories fitted to variants of the Spey are described below:-

- Shut-off cock and cold start provision

  Situated downstream of the LP governor, the shut-off cock handled both main and pilot burner fuel flows. On shut-down, residual fuel in both manifolds was dumped overboard[1] and pump delivery flow was recirculated to pump inlet during engine rundown. Additional flow for starting under cold conditions was provided at an intermediate position of the shut-off cock lever. Alternative cold start augmentation on the Phantom Spey was provided by a solenoid-controlled fuel valve. In either case, the extra fuel was switched off when the engine reached idle speed.

[1] On the Phantom Spey a 'dump' tank was provided to avoid fuel from both the engine and reheat system – a considerable quantity – being dumped on to the carrier deck, which would have resulted in a fire and safety-of-footing hazard. The tank contents were sucked into the jet pipe by an ejector nozzle following takeoff.

- Engine stall fuel dipping

  On the Allison TF41 Spey engine a device was fitted to counter engine stall due to ingestion of gun-firing exhaust gases. A solenoid valve linked to the gun-firing circuit bleeds off a large quantity of engine fuel. After gun-firing the engine returned to its original power level.

- Boundary layer control sensing and reset (Buccaneer and Phantom)

  On selection of boundary layer control bleed air, BLC pressure was also fed to a pneumatic sensing unit. This switched a supply of P3 air to a small piston on the CASC unit, which reset the P3/P2 air orifice system and thus changed the ACU setting. On early Phantom engines, the BLC sensing unit also operated a changeover valve, which reset the pressure ratio control unit scheduling the reheat nozzle. This function was later deleted, as sticking of the change-over valve could cause complete closure of the reheat nozzle and major engine surge.

- Reheat light-up sequence and reheat flow control

  On selection of reheat, a supply of HP engine fuel flow was fed to a mechanically-timed valve and then to a small injector in the exhaust stream where it impinged on a platinum element which ignited it catalytically. This in turn ignited the vapour gutter burners and thence the main reheat manifolds.

  A hydraulically-damped piston supplied with HP engine fuel gradually opened a valve to allow reheat manifold fuel to increase as the throttle was advanced in the reheat range. A far more detailed description of the reheat control fuel system for the Spey is given in Cyril Elliott's excellent book on reheat in the Rolls-Royce Heritage Trust Technical Series No 5.

- Control system diagrams

  Excellent control system diagrams were provided as usual by the Illustrations Department. These reached their zenith on the Phantom Spey, which featured all of the above accessories, plus separate reheat fuel supply and nozzle pumps and controls catering for three types of boundary bleed. The complexity and length of these magnificent drawings exceeded by far anything that had gone before and were a constant source of admiration, particularly to those Americans who saw them either at St Louis or Edwards AFB.

- A retrospective view of the efficacy of the changes introduced on the CASC

  As mentioned in the main text, one of the supposed virtues of the CASC was the metering of the whole fuel supply thus removing the need for half-ball valves and small diameter metering valves as used in the control units for the Avon and Dart, which were susceptible to blocking by dirty fuel. In the event, reliability of the CASC in service was, if anything, worse than that of the earlier units.

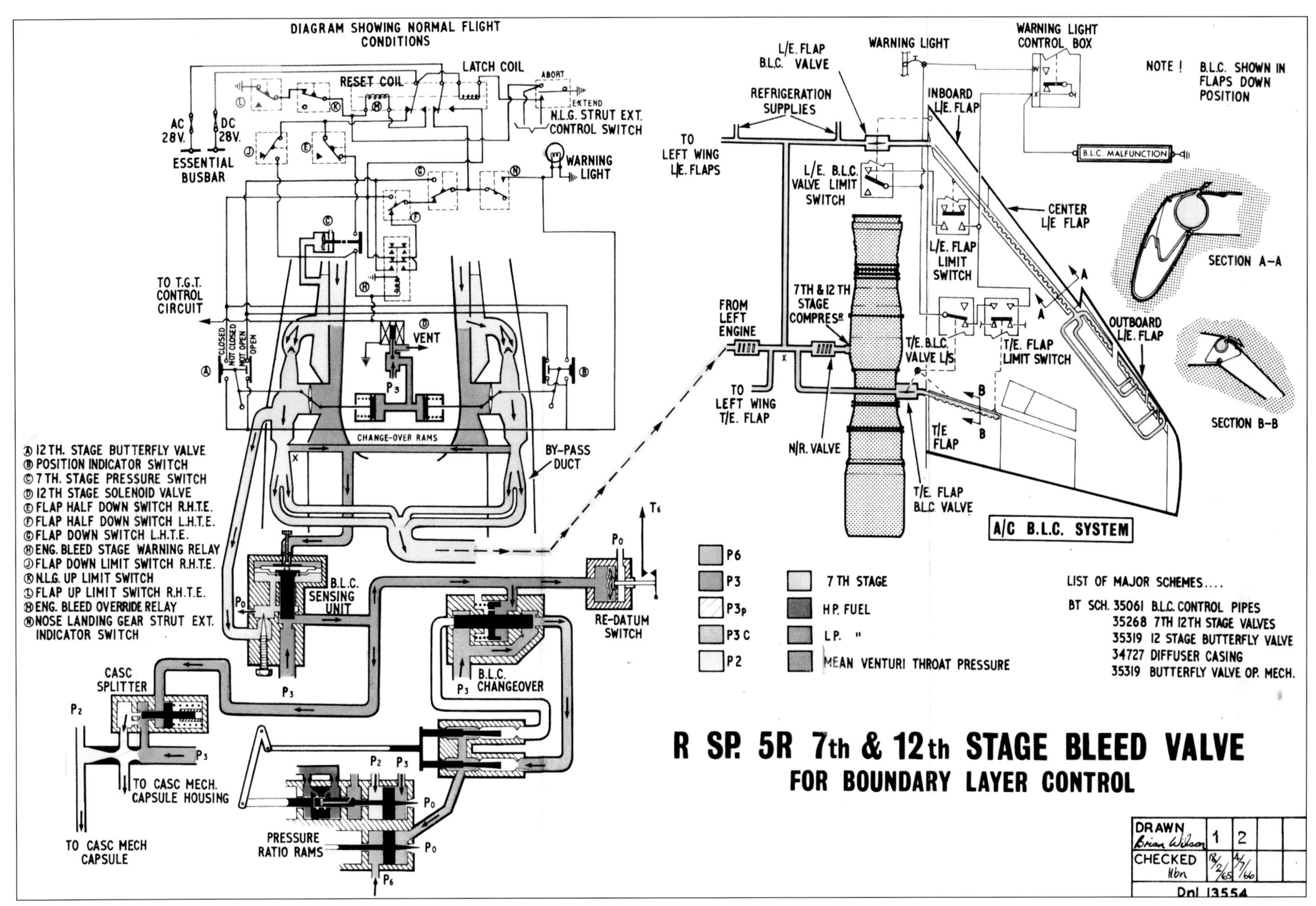

Figure 16
R SP 5R 7th & 12th Stage Bleed Valve

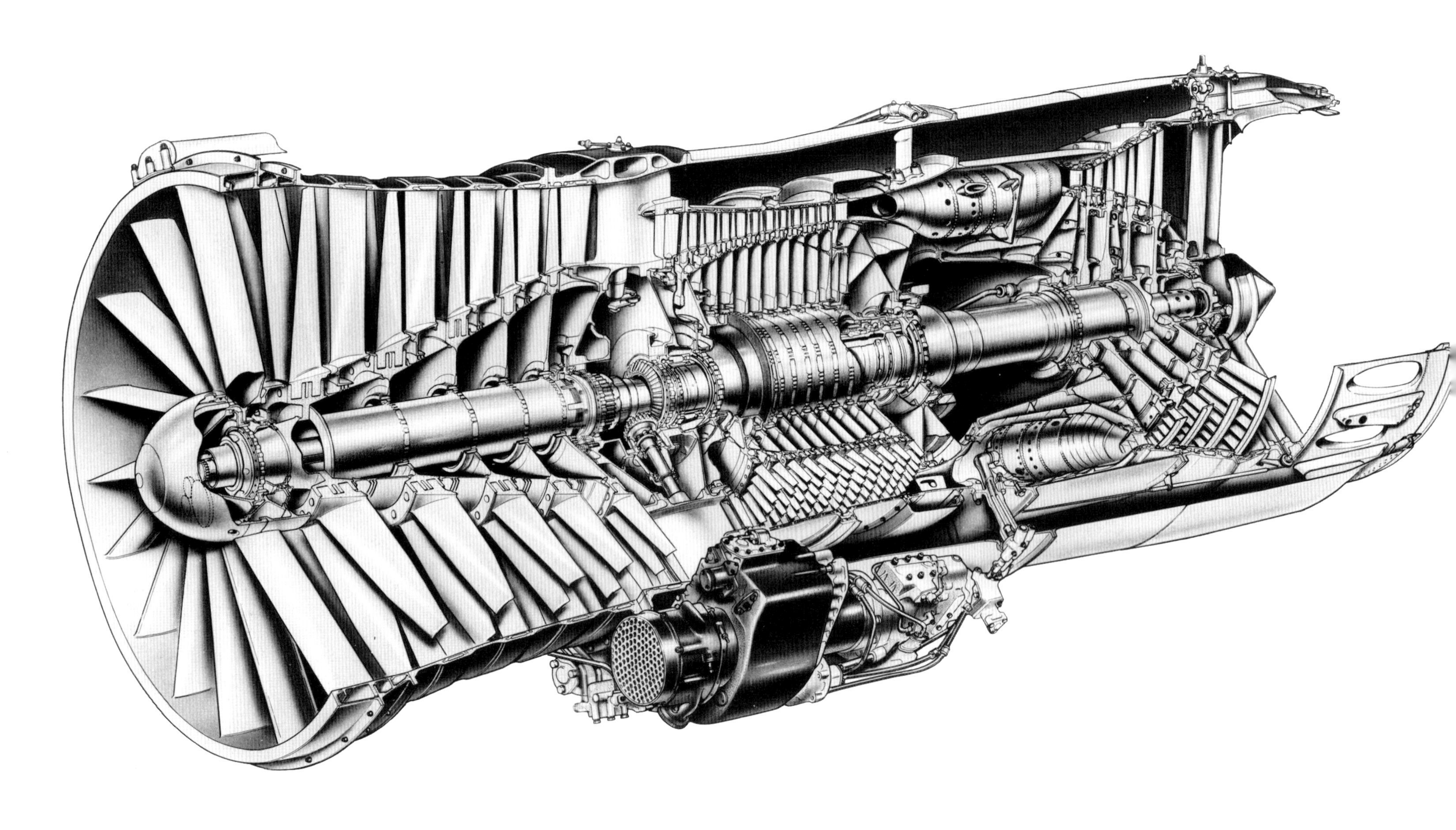

Figure 17
The RB141 - Medway

Chapter two: RB141

Medway was the name given retrospectively in 1962 to a family of engines dating back to 1957. With a bypass ratio about 0.7, they all had the same basic cycle and design configuration. Rated at between 12,000 and 15,000 lb, the RB141 was the only one which reached the hardware stage. Other projected versions were the RB140, 142R and 174, involving different scales and ratings for civil and military installations, some with reheat; none of them went into service.

The 13,790 lb RB141-3 was specified for the original DH121, for which the Government-owned BEA had placed an order for 24 with options for 12 more. By the time the engine went to test at the end of 1959, BEA had reviewed their forecasts of future market growth and reduced the size of the aircraft. Design of the smaller RB163 was already under way for a 15% scaled-down DH121 carrying up to 97 passengers. The development programme of the RB141 proceeded at the –11 rating of 15,000 lb, with an advanced Caravelle in mind.

Development testing

The original development plan was for 12 engines, which had reduced to nine following the scaling down of the DH121. Previous rig testing had already shown that the original four-stage LP compressor was drastically short of surge margin. A 5th stage was hastily added but was not available for the first three engines, which meant that handing of them would not be possible.

When engine No 1 first ran in November 1959, the LP working line was 0.2 pressure ratios higher than expected, and impending surge restricted the engine to 7400 lb. A large increase in final nozzle failed to lower the working line, and it was concluded that the mixer was choked. Increasing area by removal of a soleplate in the bypass duct lowered the working line slightly, but the engine would not run up through surge. The high working line could have been caused by excessive bypass duct pressure loss, but this was thought highly unlikely as the predicted mean mach numbers in it were so low. A very low mixer discharge coefficient was thought to be the problem.

However, when bypass duct total pressure rakes were fitted they revealed a duct pressure loss of up to 23% instead of the predicted 5%. Airflow testing on an engine carcase confirmed this measurement, and systematic removal of obstacles identified one local hump as the main culprit. This led to the fitment of fairings and by engine No 8 a new bypass duct of 15% larger area. The loss reverted to 5%, which improved the SFC by 1¼% and also allowed effective rematching which, in conjunction with changes to NGV areas, gave a further benefit.

BEA had never bought the Avon-powered Caravelle and, by mid 1960, with no RB141 production order in sight, development was confined to work which could help the RB163, which was to be a very tight programme. Until the RB163 took over, five RB141 engines carried out endurance Altitude Test Facility (ATF) and Hucknall noise bed running. Major development objectives were improving performance, reducing roughness and preventing compressor blade failures.

Other mechanical defects were encountered. There were a number of LP and HP compressor blade failures in the lugs and in the blade form. Strain gauge tests on engine and rig, and static fatigue testing enabled blade cut backs, and modifications to pin fixings to be applied to the RB141 and later the RB163. Improvements were made to leaking turbine blade locking plates. LP system roughness led to a design scheme for a spring-mounted LP turbine bearing on the RB163.

Steady improvements were made to the performance by minor modifications until a level within range of the guarantees was achieved by the end of its short career. Compressors were 2% efficiency below target, cancelled out by the turbines which were 2% above. There was still something to come from the mixer. This had 20 short chutes of rectangular section, which injected the bypass air into the hot stream. Engine and rig measurements suggested that, although the temperature mixing was good, the radial component of momentum was lost. It was thought that long chutes with axial exit would avoid this loss and still give the good mixing.

In June 1963, two years after testing had finished, a performance demonstration was run on engine 7 in support of a bid for the HS681 installation. This was a STOL high-wing aircraft with blown flaps and control surfaces requiring four 17,300 lb engines in under-wing pods, with thrust deflection. Engine 7 was built with many good aerodynamic and sealing features, large bypass duct and annular mixer. On a hot day, running at its nominal TET of 1460°K, it demonstrated 17,500 lb on the dial. It provided 20 lb/sec air offtake compared with the requirement of 12 lb/sec, and the SFC was nearly 2% better than the previous best. Needless to say this project came to nothing.

In 1960 a reheat research programme was carried out at the Allison plant in Indianapolis using one of the RB141s and a 44¼" diameter pipe. This was directed at the AR968 for the TFX project. The aim was to achieve maximum boost by burning hot and cold streams, and also good burning stability. Many mixer and gutter configurations and fuel distributions were investigated. The results were very satisfactory, and although the TF30 was chosen for the TFX (F111), there were to be later benefits to the RB168-25R design for the Phantom. A full description of this work including gutter designs and a wonderful shot of the engine and pipe at full reheat are shown in Cyril Elliott's excellent book on reheat, *Fast Jets (page 81)*.

Experience of benefit to RB163

The RB141 was a private venture without government support. Up to the end of 1960, the cost of the programme covering 9 engines and 1100 hours running was £4½ million. It was to save much time and money on the RB163.

The RB163 would have the right number of compressor stages from the start, have a large bypass duct with fairings, and a deep chuted mixer with elliptical cross-section duct outlets and axial exit. It was, therefore, expected that achieving performance objectives would be relatively straightforward. This was not to be so.

Chapter three: RB163-1

Following BEA's pessimistic forecast of the market (a view not universally shared) in 1959, the DH121 (later named Trident) was reduced in size by about 15% to carry a maximum of 97 passengers at a maximum takeoff weight of 105,000 lb. The resulting RB163-1 at 9850 lb takeoff thrust was not just a scaled down RB141 in two major respects.

First the bypass ratio was increased to 1.0 as a result of further optimisation studies, in which it was recognised by the Weight Department that the weight penalty for going to a higher bypass ratio was not as severe as had been previously assumed, largely due to weight reductions in the LP system components (see below). Secondly, noise was becoming increasingly important, even though it was many years before it was a certification requirement. An improvement in noise was achieved by the increase in bypass ratio from 0.7 to 1.0 with corresponding reduction in thrust per pound of air, and hence jet velocity. At a later stage to further reduce noise levels on both the Trident and BAC One-Eleven aircraft six lobe-silencing nozzles were fitted to the two-side fuselage-mounted engines.

To raise the bypass ratio, the HP compressor was scaled down more than the LP and, in order to keep the mixing pressures about equal, the number of LP/HP stages changed from 5/11 to 4/12. The turbine annulus was not a direct scale, due to the higher proportion of work in the HP requiring a larger area at LP inlet and an increase in LP diameter (with a steep outer hade angle) to gain some blade speed and compensate for the relatively lower NL.

The RB141 was considered to be heavy, so many lightening features were introduced, mainly in the compressors, auxiliaries and support frames:-

- LP compressor changed from shaft and discs to drum construction. Blade aspect ratios were increased by about 10%, mostly in the rotors.
- Swan neck duct 1" shorter.
- HP compressor rotors in stages 4-12 changed from pin fixing to dovetail (using Allison service experience), and material changed from aluminium to titanium. Some aspect ratios increased slightly.
- HP compressor exit diffuser included angle increased from 9° to 14°, together with a slightly shorter combustion chamber - 1½" length was saved.
- Research on model turbines with up to 20% fewer blades, led to a new space/chord correlation with a limiting lift coefficient (Zweifel). It was decided to retain the high space/chord ratios used on the RB141, but the aspect ratio of some blades was increased.

The result was a considerable improvement in thrust/weight relative to the RB141 in spite of the higher bypass ratio. A price had to be paid for meeting BEA's stringent requirements; there was no deliberate stretch built into the RB163, it just met the bare requirement of 9850 lb thrust, flat rated to only 21°C. (That did not stop de Havilland requesting an additional 250 lb takeoff thrust only a week after they had received the initial performance data! See appendix).

Development and flight testing

To make up for lost time, the design and development had to be done quickly. Design started in mid 1959 (see section in Appendix on 'A period of intense activity'), and the first development engine ran on New Year's Eve 1960. The large gathering for this major event went home disappointed because running was not possible with the bleed valve shut. This time it was HP compressor surge caused by an accumulation of poor component efficiencies, an unaccountably worse surge line than the RB141, and the compressor being designed for 3% more capacity than the engine was matched for (3% for the AR963, which never materialised). Future tests were run with cut back HP1 and LP1 NGVs to increase capacities and enable the engine to run up to design speed, but this

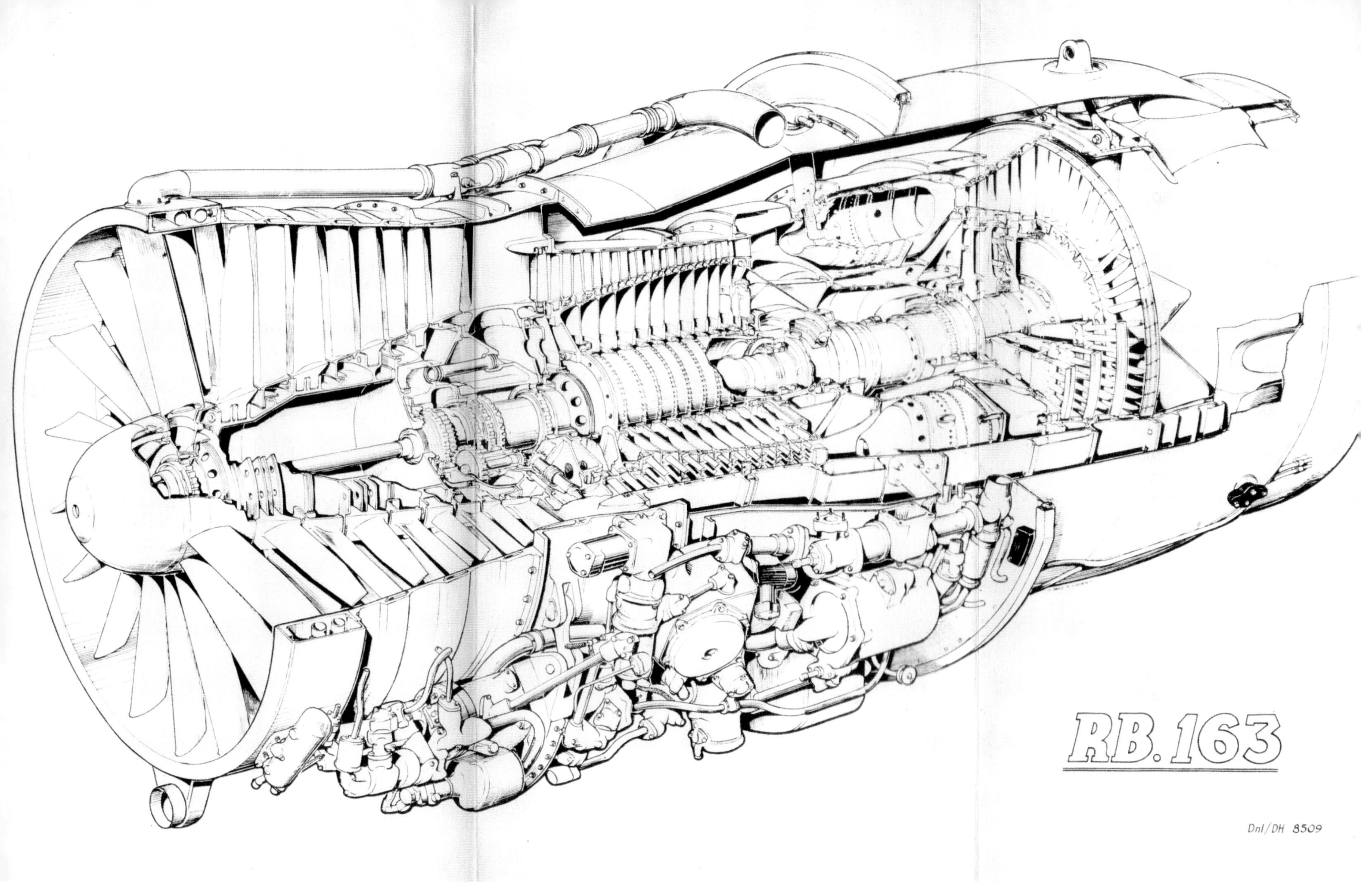

Figure 18
RB163

marked the beginning of a history of recurring HP compressor surge and overheating on all marks of Spey. The LP had sufficient margin to allow raising its working line.

At the start, the 7th stage handling bleed valve consisted of a thin metal strap like a giant jubilee clip, operated by a scissor-like toggle mechanism. This was unstable and failed after a short amount of running. In addition, it could only be operated as a bang-bang system, which was not acceptable as it gave a large jump of thrust from 4000 to 2500 lb on approach to landing. Testing was delayed until a modification could be designed. It was replaced by a thicker sliding ring, which was pushed axially over two flanges between which were the circumferential bleed slots; it could be operated progressively as it was linked with the variable IGV mechanism. It also had mechanical problems, but they were eventually cured (Figure 21, top right hand corner).

The test bed SFC of the first successful run was 18% worse than brochure, although the cruise deficit was not as great. This was worse than the RB141 had ever been, and in those days the situation seemed critical. At one time, daily performance meetings were held at Elton Road, some very acrimonious, until things had improved. Many battles were fought; research v development headed by Harry Pearson and Ernest Eltis; rig v engine; compressors v turbines. At one meeting a new LP compressor was mercifully rejected.

In March special instrumentation was fitted at HP and LP turbine exits, which indicated an LP turbine efficiency of 78% instead of a design value of 91%. A belated rig test at engine operating conditions gave 88%; excessive yaw angles of 20° and 24° were measured out of the HP and LP turbines.

It was appreciated that bodging the NGVs was a crude and inefficient way of increasing capacity, particularly the LP1. A 'proper way to go' design was initiated. Both vanes and blades were unskewed to give decent reactions and work splits, and the numbers of HP2 blades and LP1 NGVs were eventually increased. Because a blue pencil was used to draw the thrust/SFC curve, it was called the 'Blue Turbine'.

At one meeting Stan Smith, leader of the Turbine Section in the Research Dept, dramatically announced that the Zweifel Coefficient was to blame for the poor LP turbine performance. For those who had forgotten who this gentleman or his coefficient were, he explained that the LP vanes were too widely spaced to achieve the extra deflection required, and so low efficiency and high exit whirl. Ernest Eltis observed that Zweifel was German for doubt *'a very appropriate basis for this turbine design'*. Some of us doubted that Herr Zweifel ever existed, or perhaps he was a convenient turbine office scapegoat or bodge factor.

The real break through came in June when engine 6 first went to test without the increased area LP1 NGVs (+4½% instead of +14%) which were not available, and fortuitously had 7% better SFC than the average of the previous five. The penny dropped; the LP1 NGVs had been stalled all the time. As we would afterwards explain it all – a combination of HP and LP turbines poorly matched by the change of bypass ratio, fitted with vane and blade numbers which pre-supposed efficient running. Bodging open the LP1 NGVs had led to a minimal flow acceleration across these highly loaded vanes, aggravated by the large tip hade angle. They could not turn the air through the necessary deflection to meet the LP1 rotor blade entry angle. Hence a lousy LP1 stage performance, and also spoiling of the LP2 and increased jet pipe pressure loss.

Misplaced confidence had meant that no model turbine rig tests were run before the engine. They would not, however, have revealed the full extent of the disaster, because the LP rig had a parallel-sided inlet annulus, and the LP1 NGV an axial inlet angle, giving superior contraction than on the engine. Something good was to come out of this sorry saga to prevent a recurrence in the future:-

1. Testing of engine parts on the rig. Also a two-shaft model rig capable of testing both turbines in series to check for interference of HP and LP.

2. Rigorous check of NGV tip velocity ratios, and blade radius of curvature as well as Zweifel coefficient.

3. Blade and vane space/chord margins to cope with less than design engine efficiencies, and alternative sets of properly machined blades available.

Thereafter, the performance improved with the blue turbine, which came in instalments in the next three months, and a large number of minor clearance and polishing 'brasso modifications', mostly in the turbines and mixer. These included profiled turbine tip seals, cambered exhaust unit strut fairings, and cut back mixer chutes. The last one was surprising as it conflicted with RB141 and rig experience. By the time the engine went into service in 1964 it had been sufficiently developed to meet thrust and SFC guarantees. In getting there, the LP compressor, bypass duct and mixer had to be redesigned on the RB141, and now the two turbines on the RB163-1. Our consolation was that if it had been right first time, we would have been accused of putting padding in the original bid! Plots showing the performance development of both the RB141 and the Spey RB163-1 are shown in Figures 19 and 20. These plots dramatically illustrate the problems that beset the achievement of engine performance at that time and also presage the even more severe problems encountered on the RB211 only a few years later.

Bulbous tipped LP and HP compressor rotor blades were tried as an alternative means of adjusting frequency to avoid lug and blade failures. They were not successful, so we reverted to thickening and cutting back. In the case of LP1, a dog leg twist was applied (-5° in the middle) to restore the work. Eventually, new Speys had blade numbers reduced from 23 to 20. A change of pin clearance fixed LP4 resonance, but the problem was to reappear on the –25 and receive rather different treatment.

Carboning-up of burner pintles caused core burning, which burned the NGVs. The cure was frequent washing of the burners. Failures of the HP location bearing were due to excessive load, which had to be lowered by altering the diameter of the seal on the rear of the 12th HP disc.

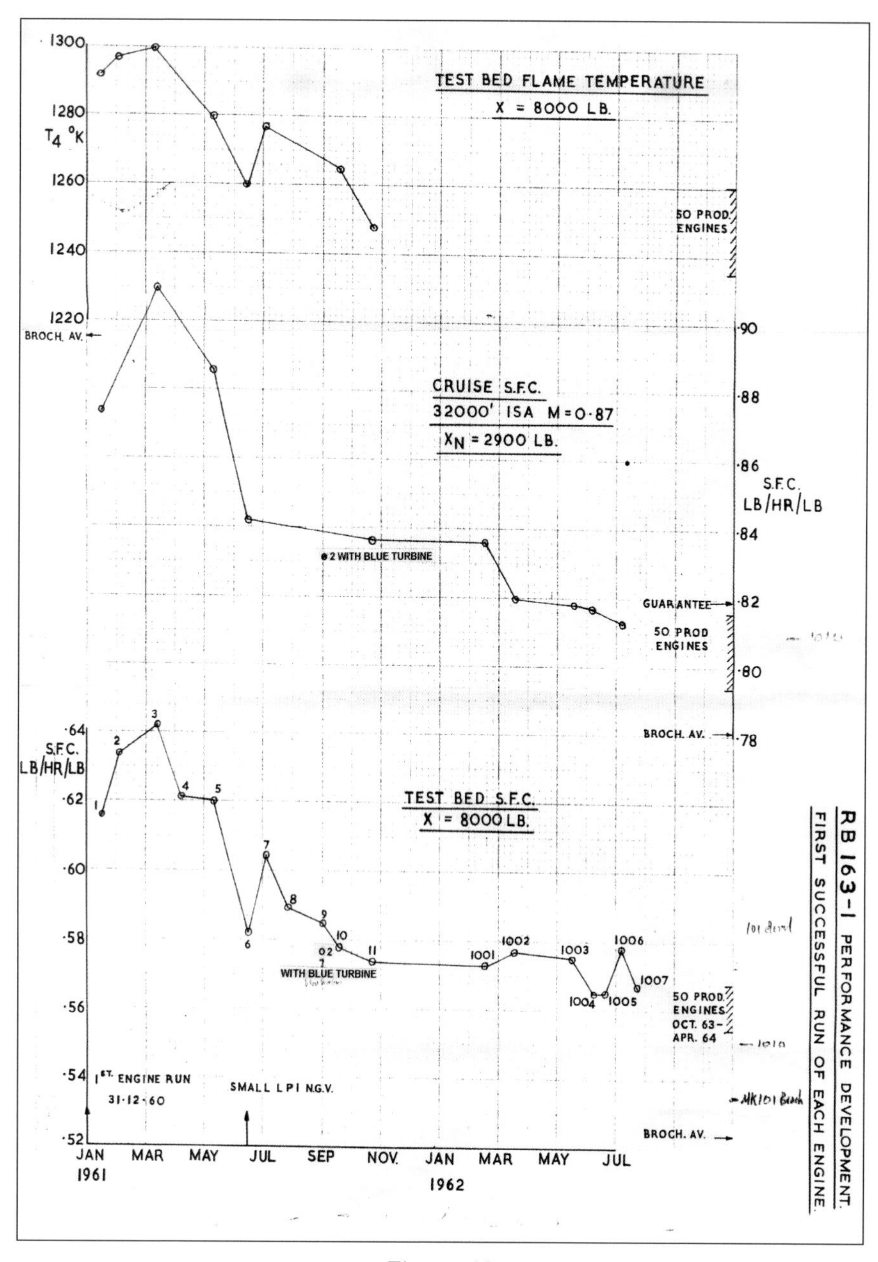

Figure 19

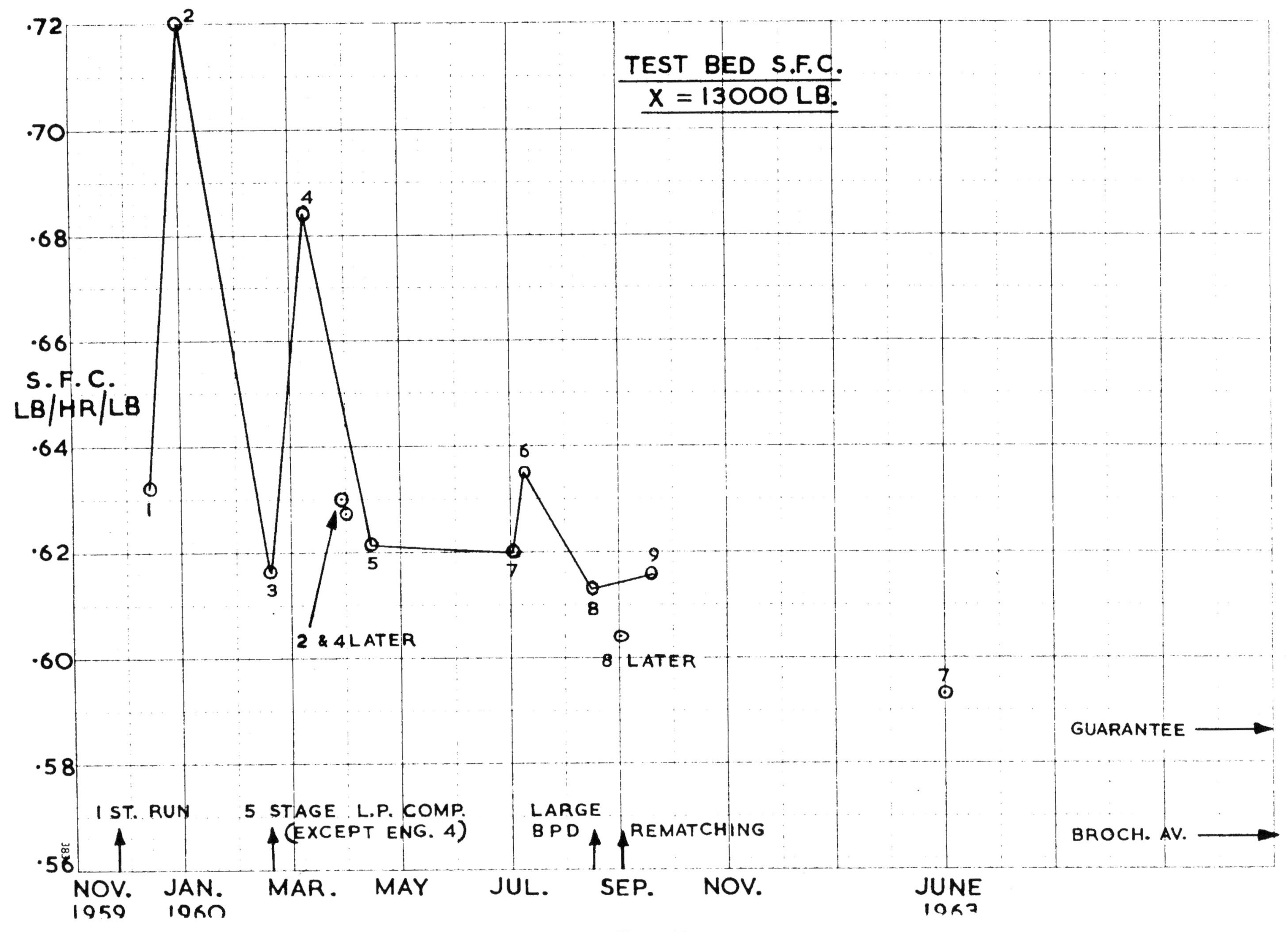

Figure 20
RB141 Performance Development
First Successful Run of Each Engine

As part of the safety features of the engine, it was considered necessary to have an emergency fuel shut-off mechanism, which would be activated in the event of a failure of the LP shaft thus preventing the LP turbine discs over-speeding to bursting point. A special test was devised to check this mechanism in which the LP shaft torque was taken by a modified quill shaft at the rear of the LP compressor, designed to break at pre-determined speeds; to begin with at idling, and then thickened up in stages to fail at top speed. In addition, a thrust bearing was introduced to stop the disconnected LP compressor from shooting out of the front of the engine. On the critical last run, attended by a selected audience, Chris Webber hid behind some filing cabinets in the control room. He needn't have worried because the fuel was successfully shut-off, and the discs did not burst. The operating cable finished up broken, possibly after the strain of shutting off the fuel! The device has never had to operate in anger. Many years later there was one example in service of an LP compressor shaft failure on the run-up prior to takeoff when the mechanism did not operate, because it also had been damaged. The LP turbine was completely free and located, and surge probably saved the turbine from excessive acceleration.

The original plan was to carry out flight testing in the Buccaneer NA 39, but it was cheaper and earlier to fit two Speys in the inboard positions of the Vulcan at Hucknall in 1961. A number of handling and control problems were revealed – sub idle NH, NH creep, failure to re-accelerate quickly above 20,000' with offtakes, TGT overshoot, deceleration surge, and – on one occasion – the Vulcan had to return on two Conways after the Speys stalled. HP compressor surges and subsequent failure of the HP3 stators was a recurring problem. IGVs and bleed valve were rescheduled to reach a compromise between handling and performance. The IGVs travelled from +30° at low speeds to -5° at high, and the bleed valve closed progressively to be shut at +10° - a revision from the angles shown in Figure 21.

Testing in the Trident began at Hatfield in 1962. The centre engine surged on a ground run with John Blackie aboard. The cures were to fit vortex generators in the intake and do another bleed and whirl adjustment. In summer 1963, Trident G-ARPC went to Torrejon base in Madrid for performance tropical trials. Instead of passengers, it was loaded with lead. It soon became clear that takeoffs with one engine cut at V1 were using too much runway. Something was wrong and fingers were pointed at the outnumbered Rolls-Royce performance man (the author). For the rest of the day, I had to stand up at the front with the crew checking the thrust of two engines as the aircraft lurched down the runway. This was not easy with a slide rule and P7-po gauge, especially when some lead broke loose and slid across the floor. I had a worry that a faulty T1 canister might move the IGVs off their negative stop, resulting in a low thrust when NH governed. As the day got hotter, more and more runway was used up on each takeoff trial until finally even 13,000' was not enough. As we approached the very end of the runway, John 'Cat's Eyes' Cunningham, in command, slowly raised his hand to have the third engine opened up from idle, and we staggered off in a cloud of dust over the Madrid sewage works. Eventually it turned out that there was a position error on the aircraft speed probe such that the indicated airspeed fell by 3½% when the aircraft was rotated on the ground at constant true airspeed. When combined with a low acceleration with one engine out, the impression given to the pilot by the ASI was of a lack of performance. It appeared that rotation was being checked without quite reaching the unstuck attitude, the prime cause of the long distance. The trials were successfully completed after the instrument had been corrected back at Hatfield. The Trident entered service in 1964.

Service problems included the expensive break up of the HP centre location bearing, requiring some strengthening of the cage, and ingestion damage to early aluminium compressor blades where thoughts were given to converting to titanium. An uncontained fatigue failure of the LP compressor of a 505F took place at Rome when opening up for takeoff. The casing disintegrated, blades were missing and the rear toilets were damaged. This was attributed to a higher number of engine cycles per flight and a wider NL range than was anticipated due to the aircraft auto-land system.

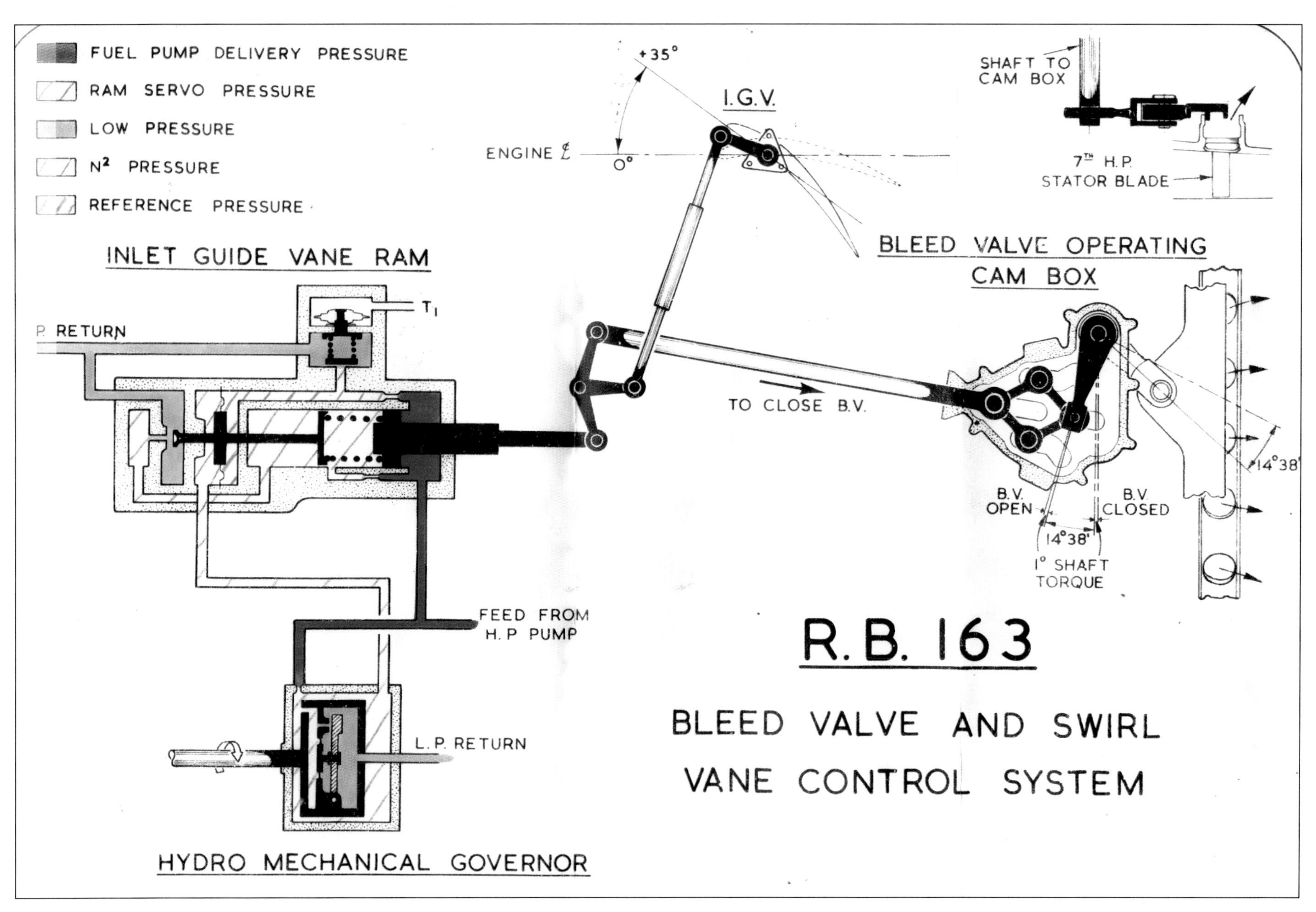

Figure 21
RB163 Bleed Valve and Swirl Vane Control System

Mk 505E & F (Trident 1)

In 1961 the Trident became the –1E with an increased takeoff weight and a longer range of 1150 nm (1710 nm at typical payload). In 1966 the Mk 505E was available with a 5% hot day takeoff contingency rating, extending the flat rated 9850 lb thrust to 29°C. It had a higher modification standard allowing a 30°C higher TET. Next year came the Mk 505F with uprated turbine materials and cooling giving a further 40°C TET. It offered a takeoff thrust of 10,050 lb, with a 5% contingency flat rated to 32°C. All 505 and 505Es were then converted to 505Fs.

Orders for the Boeing 727 soon overtook the Trident even though it was launched over a year later. It was not long before there was a realisation that the BEA market forecast was wrong and the ill-advised shrinking of the Trident was reversed to accommodate more passengers and increase the range. The required thrust increases above the 505E and F were beyond the scope of the initial Spey cycle.

Figure 22
Trident I on Takeoff

AR963 competition for Boeing 727

Two years behind the original DH121, Boeing designed their own larger three-engined 727. Rolls-Royce and Allison submitted the AR963-6 at 12,170 lb thrust flat rated to 25°C. It was a 1.21 area scale (1.1 linear) of the RB163-1, making use of the 3% larger capacity HP compressor. An uprated AR963-9 was also proposed, having a 3% larger LP compressor as well, and higher flame temperature. In 1961 the Boeing Project Design Office under Jack Steiner made this engine their prime offering for the new aircraft and, after a presentation by Boeing and Rolls-Royce, United Airlines accepted the AR963, but Eddie Rickenbacker on behalf of Eastern rejected it in favour of the alternative US engine, reportedly for personal reasons rather than technical. This turned out to be a very important decision; the P&W JT8D was created, and its subsequent success is well known.

Figure 23
The Allison/Rolls-Royce AR963

The large turbine and the McCarthy engine

After the 163-1, the performance struggle was not over. It was going to be necessary to improve and uprate the Spey quickly, whilst fighting an uphill battle against the larger JT8D. The first priority was to have another look at the LP turbine. The blue turbine had been the best of a bad job, but the LP was still low on efficiency, and had little scope for future increases in capacity and work, which were sure to be needed. In 1960, when the first Spey was being assembled, Derek Taulbut was perturbed by the amount of daylight visible through the LP1 NGV at the tips. In 1962, as turbine section leader, he was responsible for a major aerodynamic redesign. The HP was opened up in the annulus area at the back to reduce velocity into the LP1 NGVs, and the LP was a completely new design of slightly larger tip diameter. This new design of LP turbine gave the best efficiency ever achieved, and this was capable of being maintained at 10% and even 20% increases in flow.

Every component was looked at for potential 'brasso modifications', and a list of 94 drawn up for a bumper test. The test incorporated the new large turbine, but of course many of the features would be applicable to blue turbine engines. Denis McCarthy was in charge of the exercise in 1965, and was honoured to have Spey 6024 called the McCarthy engine. None of the modifications was mechanically impossible, although the cost of some was open to question. The McCarthy engine had 88 modifications including:-

- Thinned LP1 compressor clapper blades
- Optimum IGV schedule
- Abradable HP compressor linings
- Gold-plated combustion outer casing to reduce heat loss
- Honeycomb turbine tip seals
- Cambered exhaust unit fairings
- Super slim rear bypass duct

The effect of these was to improve component efficiencies, reduce internal cooling and leakage flows, and to raise compressor surge lines. Several tests were run with different matching to exploit the HP surge line improvement for maximum SFC benefit, while still meeting handling requirements. The eventual analysis of SFC improvement was:-

Component efficiencies	1.45%
Reduced cooling air etc	0.40%
Matching	2.15%
Total	4.00% SFC

With this SFC improvement went some increase of thrust at a TET, although it would be possible to swap SFC for more thrust by different matching. A total 4% from 88 modifications worked out at 0.045% each, a unit known as one 'Elt'. This approach led to a temptation to assume they were all of equal value, and could be added together like pounds of tea.

Chapter four: Four-stage LP compressor developments

Launching the RB141 and then switching to the RB163 on an order for 24 Tridents had been a bold step, but to justify this decision more homes for the engine needed to be found quickly. To begin with, it was not going to be possible to offer significantly more thrust as the RB163-1 had no deliberate growth built in and, although its targets had been met, there were no margins. Larger upratings would have to wait until operating experience at the basic level had been attained, or advances in technology made.

Engine improvements available soon after the –1 without drastic structural and mechanical changes would be the more efficient large turbine, slight adjustment of ratings using upgraded turbine materials, and water injection. In addition, there would need to be a few changes of material and thicknesses to blades, discs and casings, etc, especially for any military versions.

RB163-2 and 2W (BAC One-Eleven 200)

Figure 24
The BAC One-Eleven at East Midlands Airport

In 1961 the two-engined BAC One-Eleven was announced. It was based on the H107, an earlier design by Hunting Aircraft, and was clearly sized for the committed Spey. Initially, it carried up to 79 passengers with a maximum takeoff weight of 78,500 lb, and flew at up to Mn, 0.78, considerably slower than the Trident. The BAC One-Eleven bore the same relationship to the DC9 and Boeing 737 as the Trident did to the 727; in both cases, the smaller size was to be a handicap to future sales.

The BAC One-Eleven required relatively more takeoff thrust than the Trident to cope with the larger reduction of thrust with an engine out. Upgrading HP1 and 2 turbine blade materials to N108 and 115 enabled a SL ISA thrust of 10,410 lb, an increase of 5½%, and an 8½% hot day increase. For the first customer, British United Airways (BUA), the hot day thrust increase was doubled by means of water injection to meet their special runway requirements; this was the RB163-2W engine. Coming slightly later was a Braniff order for the dry RB163-2. Although the dry performance was the same for both, the design was different in several respects apart from the addition of water. Design started two years after the –1, and the BAC One-Eleven entered service in 1965, one year after the Trident. This meant a tight programme especially for the –2W, which retained the –1 blue turbine, whereas the –2 had the more radical large turbine.

The system adopted for the –2W, and later 'wet' Speys, was to inject demineralised water directly into the primary zone of the flame tubes through the pre-swirlers. By injecting up to 1350 gallons per hour of water 7 to 8% thrust boost was achieved, depending on ambient pressure, with a water/air ratio of 3.7% which gave a safe margin from extinction. The fuel flow was increased by 20% to evaporate and superheat the water. The TET was limited to 10° below the dry level, and control system resets applied to governors and the ACU. Water was stored in a 125-gallon tank and injected into the engine by a turbo-pump driven by HP compressor delivery air. Any residual water after takeoff had to be drained overboard into the outlet of the toilet wash system in order to prevent it freezing. To offset the effect of a rising working line with

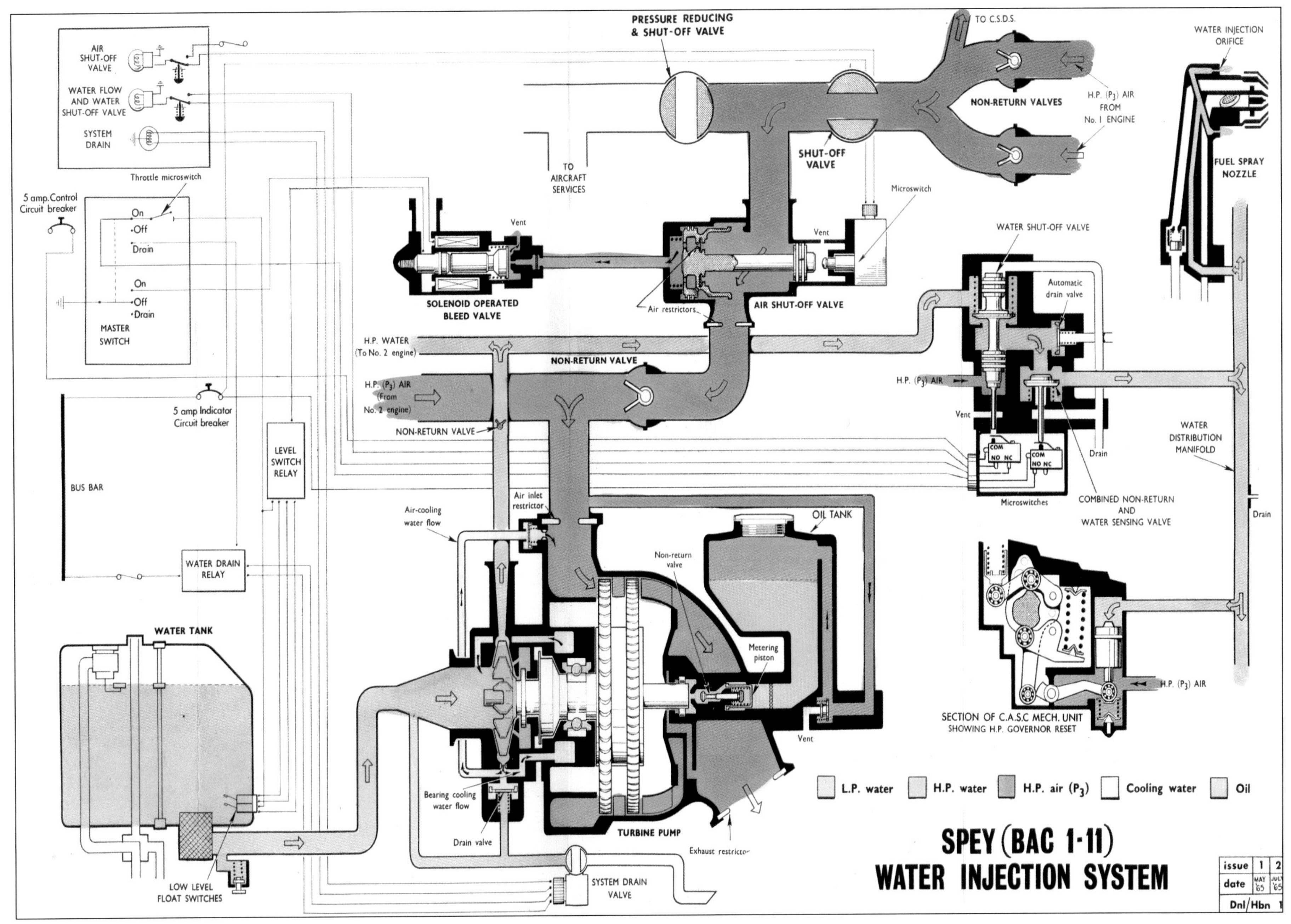

Figure 25
Spey (BAC 1-11) Water Injection System

water injection, the –2W had a 4% smaller HP compressor capacity than the –1, achieved by positive twists of front stage rotors.

With the large turbine, the –2 was able to increase the LP turbine capacity to rematch for 3% higher flow HP compressor built into the –1. This resulted in a lower TET than the –2W, and the more efficient turbine gave a lower SFC, so the –2 had the better all-round performance until the –2W recovered by fitting a number of McCarthy modifications including a 'slim-line' bypass duct, abradable linings on the LP stators, and going to fully machined HP NGVs instead of crimped. There were to be a few occasions when a special rating had to be offered to overcome a local difficulty. On the –2, a +3% hot day takeoff thrust was available to Braniff to clear an obstruction at Kansas City. In the end, this was not required, but 1½% of it was used by Mohawk on a few summer occasions.

Unlike the Trident, the BAC One-Eleven did not originally feature chuted silencing nozzles but, like the Conways had con-di nozzles to minimise the presence of shock-cells in the core of the jet, which were not only a considerable source of annoying 'tearing' noise, but also to minimise the effect on tailplane structural fatigue resulting from the same extremely intense source of sound energy. The engine was fitted with a clamshell reverser.

One year after going into service, the –2W and –2 overhaul life was around 2000 hours, with unscheduled removal rates of 0.34 to 0.41 per 1000 hours. Limiting features were HP1 NGV deterioration requiring increased cooling and a change of material, and burnt flame tube inter-connectors cured by increasing their size and cooling. Other service problems were connected with HP compressor handling. Surging resulted in blade damage from flexing. The cures were optimisation of IGV and bleed scheduling, fixing bleed valve failures and cutting back blading. Overheat from deep HP compressor stall was a general problem described in the Appendix.

In 1971 a later model of the BAC One-Eleven with more powerful Speys crash-landed onto an autobahn following a water injection takeoff. The water tank had been filled from five unmarked containers, three containing water and two kerosene. Because the water settled to bottom of the tank, below the kerosene, the engine temperatures were normal during takeoff. The aircraft was already airborne when the lighter kerosene fuel was reached causing severe overheating damage to the turbines followed by flame-out of both engines. The kerosene injection corresponded to about 150% over-fuelling, so that even if the engine fuel system had enough authority to cut off all its own fuel, the result would have been the same – severe overheat followed by surge, rich extinction and complete loss of thrust.

Figure 26
A Spey RB163 being installed in a BAC One-Eleven – the thrust reverser operating mechanism is clearly shown

RB168-1 Mk 101 (Buccaneer)

One good result of the reduced thrust requirement for the DH121 in 1959 was that the resulting Spey was the ideal size for the Hawker Siddeley Buccaneer Mk 2 naval strike aircraft, designed for operation from a carrier and flight at near sonic speeds at sea level. Formerly the Blackburn NA39, the Mk 1 was powered by two de Havilland Gyron Junior engines of 7100 lb thrust. The Spey Mk 101 specified for the Mk 2, provided greatly increased thrust all round, and its lower fuel consumption much improved aircraft range.

The Mk 101 was effectively a military version of the civil Mk 505 (small blue turbine) with the throttle opened to give 11,030 lb thrust at a turbine entry temperature of 1400°K on an ISA day (there was no flat rating on military engines). Many changes of material, casing thickness, etc, were necessary for the more severe military duty and ratings. For example, civil Speys operated on the P3 limiter at standard day takeoff at sea level, so this had to be raised considerably to permit full power at sea level 0.85 Mach number, with an equivalent increase in T3. As well as the more severe HP compressor delivery conditions, the LP compressor casing had to be changed from magnesium to aluminium, and some turbine discs and shaft materials upgraded. In common with other military aircraft fitted with ejector seats, the LP shaft failure cut-off device was deleted to save weight.

An important feature of the Buccaneer was its use of a large quantity of engine air for boundary layer control (BLC) for carrier takeoff and landing. BLC air was fed to outlet slots near the wing leading and trailing edges to lower the stalling speed of the aircraft and greatly assist with carrier operation. Although the offtake of air caused a significant loss of basic engine thrust, particularly at takeoff, the improved low speed aircraft handling and the recovery of some of the thrust in the wing trailing edge slots resulted in an overall net benefit. The 7th stage of the HP compressor was chosen for supply of the BLC air, the same station as for the compressor low speed handling bleed. At this pressure, the requirement was for 14.2 lb/second, equivalent to about 14% of core airflow, although a maximum of 18 lb/sec could be extracted – but not at the same time as handling bleed. The existing ports in the stator platforms were inadequate, so an angled platform was designed with longer stators and some changes were made to the blading upstream of the bleed slots – see Figure 27. Compressor rig tests were carried out with large quantities of bleed, up to 20%, and these showed satisfactory operation of the compressor with high bleed flow and no loss of non-bleed performance. It was calculated that with BLC in operation, the LP compressor working line fell by 0.1 pressure ratios, with consequent loss of thrust and pressure. The original 10-chuted mixer was designed with four variable flaps, which closed with BLC in operation, reducing the effective cold area from 217 to 184 sq ins to restore the working line. The flaps were supposed to close using HP compressor air, but this did not work on development test, so the closing had to be done by hand. One night a performance curve was run on the bed, nominally in BLC, but inadvertently without closing the flaps. This showed the working line dropped by only 0.7 R, and the thrust loss to be less than forecast, so the necessity for variable flaps was considered doubtful, especially as their use caused hot streaks in the jet pipe. The engine finished up with a fixed annular mixer. Such a mixer had been fitted on the first engine run, and in view of the long flight pipe it gave adequate mixing performance. Incidentally, the flight pipe gave a higher working line than the slave pipe, but the reduced surge margin was found to be acceptable.

Control system resets were necessary with BLC because of the large change in some performance characteristics; for example, the fuel flow demand increased and the ACU had to be re-datumed to allow satisfactory acceleration in BLC. There is no doubt that engineering the Spey to provide high levels of bleed for boundary layer control for the Buccaneer was of great benefit when it came to doing the same for the RB168-25R in the Phantom, though the latter was a much more complicated system than the Buccaneer.

Starting with an existing Buccaneer Mk 1, the Mk 2 installation was a bit of a squeeze to fit the larger Spey engine. This required a smaller bypass duct, a large pipe system to accommodate the BLC pipes on the aircraft, and an angled jet pipe.

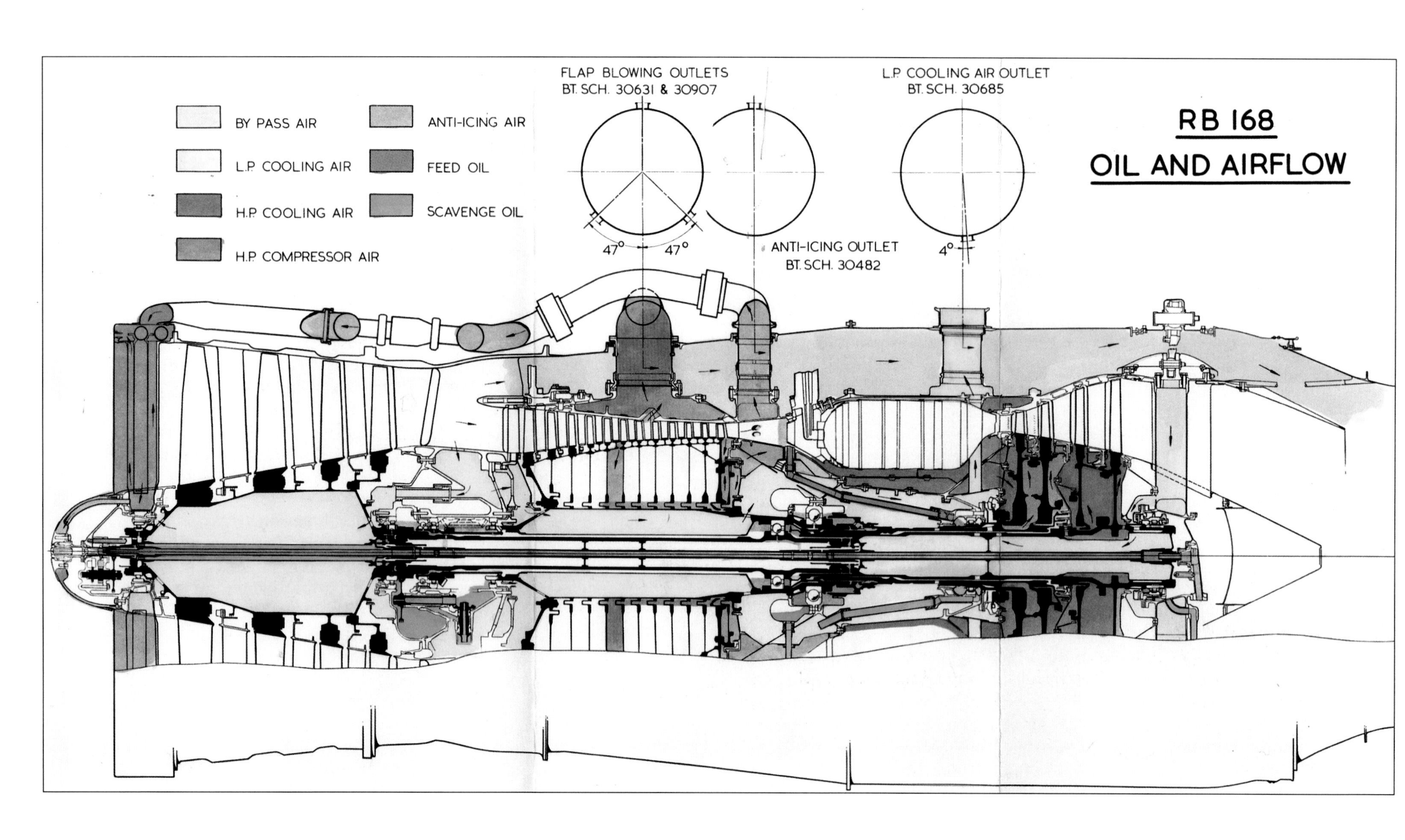

Figure 27
RB168 Oil and Airflow

Figure 28
Hawker-Siddeley Buccaneer with a Spey RB168 in the foreground

The Mk 101 came a year after the Mk 505, and the design and development programme were concurrent with the Mk 506. Even before the Mk 505 had its first run in 1960, the NA39 was proposed as a flying test bed for the Spey, but the Vulcan was finally chosen. The first Mk 101 run was on New Year's Eve 1961, exactly one year after the Mk 505. Its SFC was 3% worse than the best Mk 505, but it was lacking features such as BPD fairings and a variable chuted mixer and, of course, allowance had to be made for some loss producing features such as a smaller BPD with more blockage. Improvements had to be made to the engine to achieve guarantees. Another McCarthy type exercise was carried out with 41 performance improvement features. Many of these concerned sealing with rubber, SQ compounds or Saureisan cement to reduce leakages. Aircraft modifications were needed to the nacelle at the rear to reduce drag and to redirect the jet 5° towards the line of flight.

The handling bleed valve had to be different from the civil, which caused us some trouble on the last hour of the 150-hour type test when the strap failed in flutter. The test was allowed but a lot of effort went into engine and rig before the problem was fixed.

A major failure was the HP compressor stage 7 rotor during BLC running, which together with other similar failures delayed delivery of flight engines. A correlation was found between blade failure and low torsional frequency. Adjustments were made to the inlet angles of blades in front end stages, as this was a factor in the Compressor Office frequency parameter. This enabled the compressor to accept the large bleed.

In common with other types of Spey, the Mk 1 suffered from 'overheat' especially during a hot reslam. The engine stagnated and as a result the HP1 and 2 turbine blades were burnt out. This phenomenon will be discussed in greater detail in the Appendix.

To the surprise of some, the Mk 2 flew inverted on its first flight test – deliberately! Service started in 1965 with the Royal Navy and later with the Royal Air Force. They were finally retired in 1994. The Buccaneer S Mk 50 for the South African Air Force was fitted with two Spey Mk 101s, and also one Rolls-Royce Bristol BS605 twin chamber retractable rocket engine in the rear fuselage. This engine, rated at 8000 lb for 30 seconds, was used to boost takeoff from hot and high airfields.

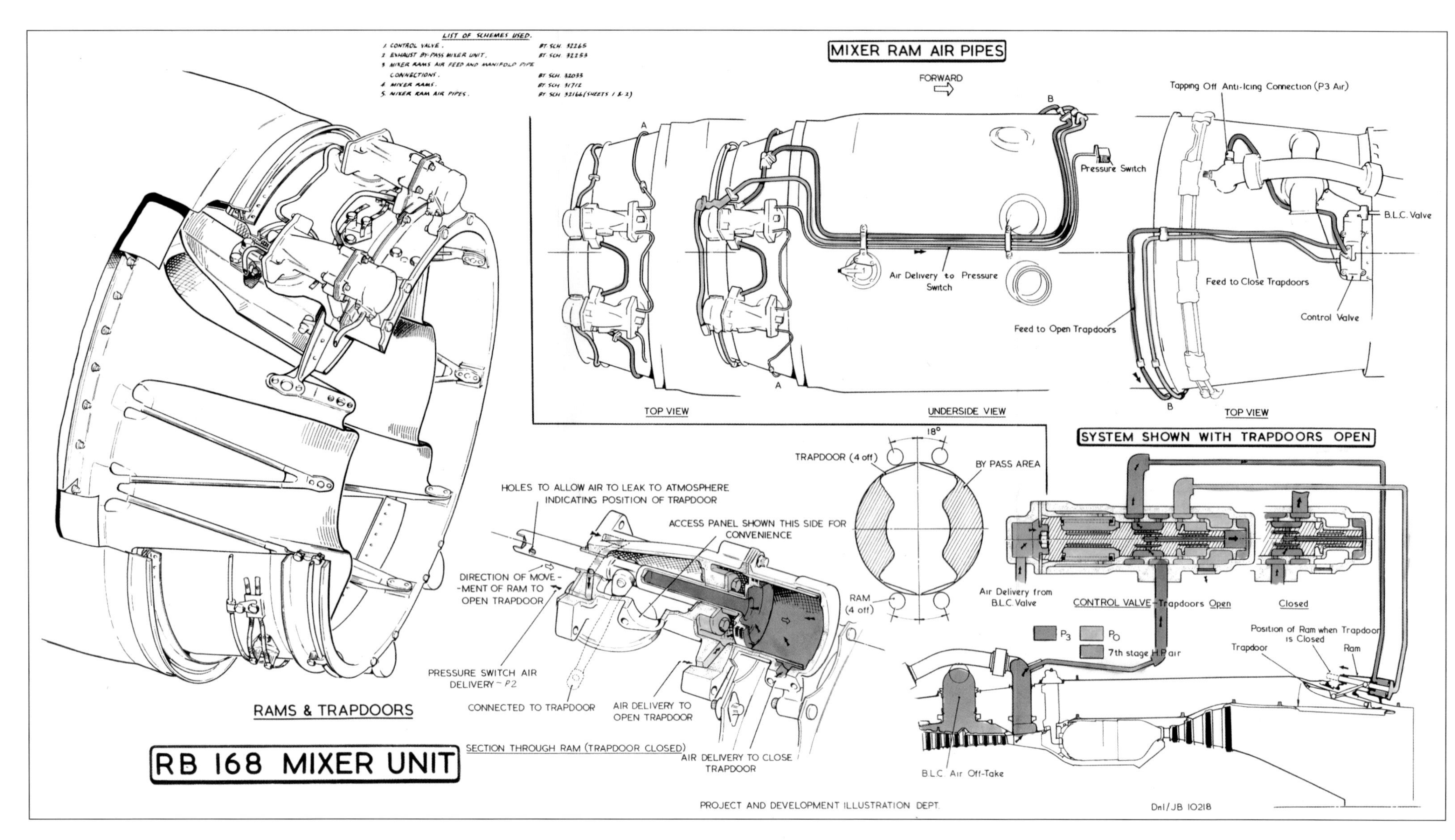

Figure 29
Proposed RB 168 Mixer Unit (not used)

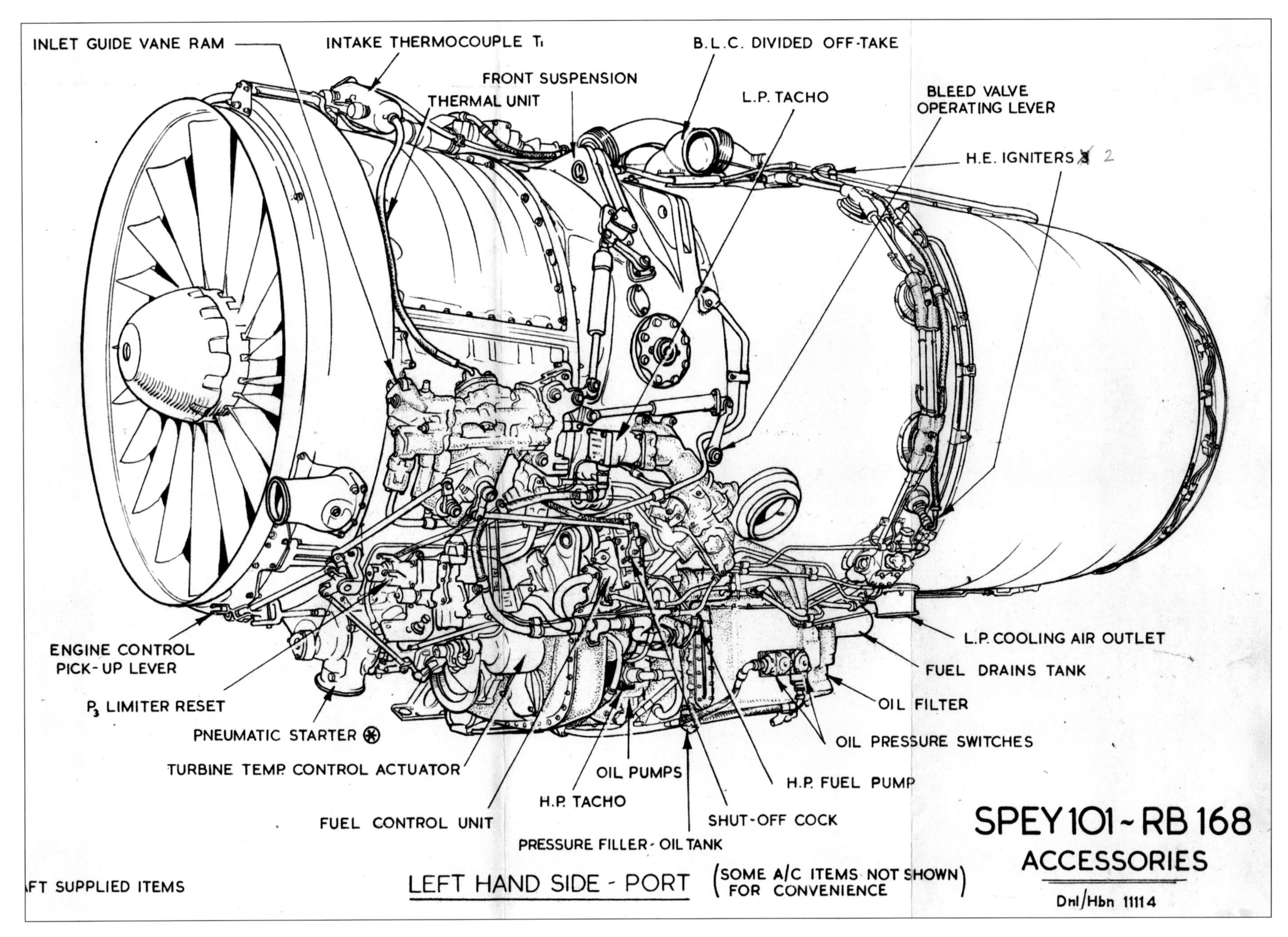

Figure 30
Spey 101 - Accessories

RB168-1 Mk 807 (AMX International)

Figure 31
Italian/Brazilian collaborative AMX military aircraft

Many years later than the Mk 101, the Spey Mk 807 was chosen from a long list of alternatives to power the single-engined AMX. This was a single-seat aircraft for close support, interdiction and reconnaissance use for the Italian and Brazilian Air Forces and manufactured by Aermacchi and Embraer in partnership. The Mk 807 is similar to the Mk 101 with the same performance, but with BLC deleted, 1½ feet shorter and 54 lb lighter achieved by using structural parts from the Mk 555. Being a single engine, it has a back up manual fuel system. The engine is manufactured under licence by Fiat Avio, Piaggio and Alfa Romeo in Italy and CELMA in Brazil. A lash up engine test was run in 1980, the first flight in 1984, and service commenced in 1989. A two-seat version for additional use as a trainer, had its first flight in 1990.

The maximum takeoff weight of the AMX is 28,660 lb, the lightest Spey-powered aircraft. The operational weight of the aircraft empty is 14,771 lb, which is lighter than the Trent 800 powerplant.

RB183-1 & 2 (F28)

Another advantageous result of the shrinking of the Trident was that the Spey was the right size for the Fokker Fellowship F28, which followed the smaller Dart-powered F27 Friendship. First announced in 1962 and in service in 1969, the F28 carried up to 65 passengers with a maximum takeoff weight of 65,000 lb. Initially Fokker sought an engine of 7500-8500 lb takeoff thrust, and the competition was between a derated RB163-1 and a Bristol Siddeley submission. The Spey Mk 550 at 8740 lb thrust, -11 to 12% less than the RB163-1, corresponding to 70° lower TET – was eventually selected. This permitted deletion of cooling for the HP1 turbine blade and HP2 NGV, giving a further performance benefit.

Figure 32
The Fokker F28 showing a typical approach into a shortfield

The typical F28 mission at 100 to 500 nm was half the Trident range, at speeds similar to the BAC One-Eleven, using short, low strength runways and with a quick turn round and self-support capability. Even under these relatively arduous conditions, low weight, low price and overhaul costs and simplicity were needed from the engine. The engine was known as the Spey Junior Mk 550, but Fokker wanted to dissociate it from the other noisier Speys. Arthur Hare suggested calling it the SPEAN, a smaller Scottish river; anmeans small in Gaelic, and the Spean contained trout rather than the larger salmon in the Spey. In the end, we settled for the RB183 instead of the RB163.

Because it had enough pressure, engine air offtake for aircraft services was taken from stage 7 HP instead of both LP and HP delivery on other Speys. This eliminated the need for a second aircraft heat exchanger. To avoid an aerodynamic mismatch of stages fore and aft of the tapping with up to 7% of air extracted, blade angles in front of stage 7 had to be adjusted as on the RB168-1. As stage 7 was the engine handling bleed station, aircraft air was switched to HP delivery at low engine rpm.

The result was a cheaper, slightly lower SFC, quieter and 8% lighter engine than the RB163-1. However, when the first development engine went to test in 1965, Fokker requested an uprating back to the RB163-1 takeoff thrust of 9850 ISA, flat rated to 2° higher AIT. The resulting RB183-2 or Mk 555 had turbine cooling reinstated, making it similar to the RB163-2W except for water injection. The weight increased by 4%, still leaving a 4% lighter engine than the RB163-1.

A number of development and service problems arose, some of which were peculiar to the RB183, such as failures of the stage 7 bleed manifold and the fibreglass bypass duct. The latter suffered from porosity caused by blistering of the covering of intumescent paint. Costly repairs forced the return of the titanium duct. The only serious uncontained Spey engine failure occurred on a 555P (see later) in 1988, when an LP1 turbine disc oversped and burst at cruise. The disc and blades punctured the fuselage causing rapid decompression, and depositing of blades in the toilet. There had been a destructive rub between the LP1/2 inter-stage seal and the drive arms of the turbine discs. Action taken was to rework the seal geometry and dimensions.

Further developments of the 555 followed to meet the needs of stretched F28s and noise legislation. Starting with the initial F28-1000, Fokker introduced the 2, 3, 4 and 6000 with up to 85 passengers (see chart in Appendix). These involved permutations of increased length and wingspan, taking the maximum weight to 71,000 lb. Having a higher bypass ratio than the RB163-1 meant the 555 had a higher TET, with not much scope for opening the throttle. Fokker initially pressed for the Mk 512 (next chapter), which would have needed a new stage 7 bleed valve and the full noise kit to meet the increasingly tough requirements. The outcome was the 555H, which merely opened the throttle to give 9900 lb thrust, flat rated to 29.7°C instead of 22.5°. In 1972 the F28 was the only conventional jet airliner to meet the FAA noise levels for new aircraft, without the need for quietening features in its engines. It was 5-7 db quieter than the Trident, and it was thought that part of this might be due to the annular mixer. A comparative test of annular and chuted mixers at Hucknall did not confirm this. Nevertheless, FAR part 36 annex 16 led to even the 555 and 555H being fitted with a 10-lobe silencing mixer with acoustic lining in the jet pipe instead of the annular mixer; these were called the 555N and P. Modifications to the flame tubes were also introduced at a later date to reduce pollution.

As well as being the quietest Spey, the 555 had an unrivalled reliability record with an in-flight shutdown rate of only 0.25 per 10,000 hours. For this reason, 555 features were incorporated in future Spey derivatives when possible; the AMX and the Tay.

RB 163-25 VARIANTS

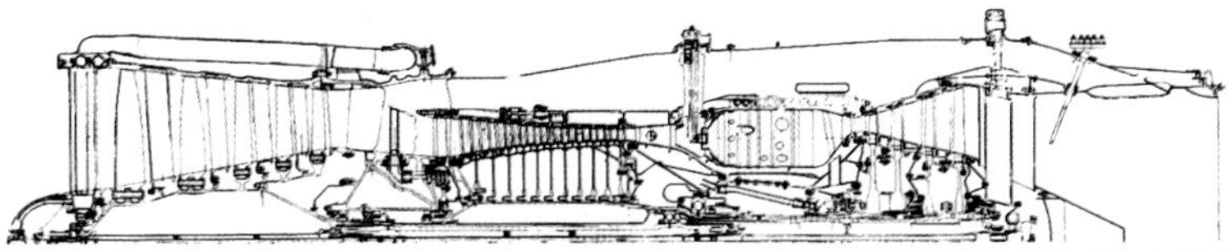

MARK NO	INSTALLATION	ENGINE DIFFERENCES	MIN T.O. RATING I.S.A. S.L.	CUSTOMER
510-14	BAC I-II AS 506-14	• ADDITIONAL STAGE L.P. COMP. • H P COMP INCREASED CAPACITY STAGE 1. • NEW CASING REQD. TO SUIT HP TURBINES. INCO 907.	11,000 lb	AMERICAN
511-14	BAC I-II AS 506-14		11,400 lb	KUWAIT
511-14W	BAC I-II AS 506-14AW	• WATER FEATURES IN ADDITION TO ABOVE.	11,400 lb WET & DRY	PHILLIPPINE
511-5W	TRIDENT AS 505-5	• WATER FEATURES IN ADDITION TO ABOVE.	11,400 lb WET & DRY	KUWAIT IRAQUI PAKISTAN
511-8	GULFSTREAM • TRIDENT TYPE H.P. & L.P. W/C, PLUS. SPECIAL ACCESSORIES. • NO FIREPROOF BULK-HEAD. • ENGINE ROTATED 13°. • MODIFIED PLUMBING. • DIFFERENT LT HARNESS • DIFFERENT FUEL PRESS SWITCH. • GRAVITY OIL FILLER NOSE COWL, JET PIPE & T/R ~ GRUMMAN SUPPLY.	AS 510-14	11,400 lb	EXECUTIVE

CUSTOMER VARIANTS

BASIC TYPE	CUSTOMER IDENTIFICATION	CUSTOMER	REMARKS
MK 505-5	MK 505-5/10	BEA	
MK 506 14	MK 506-14/10	AER LINGUS	• AS MK 506-14 BUT WITH LT HARNESS WITH CANNON CONNECTIONS. • REVERSED ROTATION HYDRAULIC PUMP.
	MK 506-14/11	MOHAWK	• REVERSED ROTATION. HYDRAULIC RUMP. • OIL TANK SIGHT GLASS IN AMERICAN PINT.
	MK 506-14/14	ALOHA	• LT HARNESS WITH CANNON CONNECTIONS. • REVERSED ROTATION HYDRAULIC PUMP.
	MK 506-14/15	BRANIFF TENNESSEE GAS	• LT HARNESS WITH CANNON CONNECTIONS. • REVERSED ROTATION HYDRAULIC PUMP. • PROVISION FOR FITTING AIRESEARCH C.S.D.S. • OIL TANK SIGHT GLASS IN AMERICAN PINTS.
	MK 506-14AW/10	HERR HORTEN	• LT HARNESS WITH CANNON CONNECTIONS. • REVERSED ROTATION HYDRAULIC PUMP.
MK 510-14	MK 510-14/10	AMERICAN AIRLINES	• REVERSED ROTATION HYDRAULIC PUMP. • FITMENT OF G.E. FLOWMETER. • OIL TANK SIGHT GLASS IN AMERICAN PINTS. • AMERICAN TYPE OIL FILLER CONNECTION.
MK 511	MK 511-14/10	KUWAIT	• INTRODUCTION OF TERMINAL CONNECTIONS ON ENGINE • HARNESS AND ACCESSORIES.
	MK 511-14/11	PAGE	• REVERSED ROTATION HYDRAULIC PUMP. • AMERICAN SIGHT GLASS.
	MK 511-14W/10	PHILIPPINE	• REVERSED ROTATION HYDRAULIC PUMP. • PROVISION FOR FITTING G.E. FLOWMETER. • FITTING OF G.E. FLOWMETER. • OIL TANK SIGHT GLASS IN AMERICAN PINTS. • INTRODUCTION OF HIGH/LOW IGNITION.
	MK 511-5W	KUWAIT	
	MK 511-5W/10	PAKISTAN INTERNATIONAL AIRLINES	• DOUBLE ELEMENT THERMOCOUPLES. • H P WHEELCASE INCREASED CAPACITY STARTER.
	MK 511-5W/11	IRAQUI	• DOUBLE ELEMENT THERMOCOUPLES.

Figure 33
RB163-25 Variants

Chapter five: Five-stage LP compressor developments

RB163-25 and RB168-20

One year after the RB163-2 came the need for a similar thrust uprating for a larger Trident and BAC One-Eleven. Having used up the available throttle opening equivalent to 40° TET on the –2, a further 8½% hot day takeoff thrust would necessitate a larger design tear up, and it was desirable that there should be potential for a further stretch at a later date, now that BEA's forecast of traffic growth had been found to be pessimistic. There were two alternatives, either to increase the engine airflow or the jet pipe pressure and hence jet velocity. For the former, a large scale-up of LP compressor would be required as all the extra flow would have to go down the bypass duct. This would have needed a larger inlet and hence a new and fatter nacelle with higher pod drag and considerable launch costs, also a much higher LP shaft torque and LP turbine loading, aggravated by the lower NL. The method eventually selected to achieve the additional thrust was to add a 5th stage to the back of the LP compressor increasing the pressure ratio and thus reducing the exit non-dimensional flow. With the same HP compressor this meant that more lb/sec flow would go through the core, reducing the BPR from 0.98 to 0.67, and increasing the jet pipe pressure. The penalty was a higher noise level resulting from the ever increasing jet velocity relative to the 163-1, which would lead to the need for more silencing features in future years.

The 5th stage on the LP compressor increased its pressure ratio from 2.19 to 2.76, a compressor similar to the RB141. Because of the higher T2, a 5% increase of HP compressor capacity was needed to avoid higher NH. This was achieved by having longer HP1 rotors and stators. The net effect was a similar surge margin to the –1, but with some non-dimensional scope for future speeding up. The HP pressure ratio fell from 7.67 to 6.91 but the overall pressure ratio rose from 16.8 to 19.1, with a 40° higher T3, and a 40% increase in LP shaft torque. With the larger turbine of the 163-2, capacities were increased efficiently to obtain the desired matching. Jet pipe mixing of the streams was satisfactory as cold/hot pressure went from slightly below one to slightly above.

The final thrust increase of 9% was achieved with a TET 15° above the –2, but initial service operation would be at 15° cooler giving 5½ per cent thrust increase. The corresponding ISA takeoff ratings were 11,400 lb (Mk 511) and 11,000 lb (Mk 510), augmented by water injection on hot days. The cruise SFC increased only slightly as the rise due to the lower bypass ratio was almost offset by an improvement due to the higher overall pressure ratio. The engine length was increased by 4½", and the weight by 70 lb on top of the 163-2, which was itself 22 lb above the 163-1.

The 5th LP rotor was overhung behind stage 4 on the rear drum. On previous marks, stage 4 rotor excitation had been a problem due to a pressure wave transmitted upstream from the 8 intercase struts via the OGVs, which coincided with the blade's natural frequency. On the –25 the OGV to strut gap was increased from 1" to 3". The pressure variation was eventually reduced to negligible proportions by cyclic variation of the OGV stagger across the pitch of the struts -5°, -2°, 0°, 0°, +2°, +5°, etc. The extra number of parts was not popular with manufacturing.

Important parameters affecting the mechanical changes on upratings were:-

1. HP compressor delivery temperature, T3 in conjunction with NH and P3. The increases required an upgrading of blade and disc materials rearwards from aluminium to steel and then to nickel. Nickel being very costly, civil engines used steel discs with angled struts between stages 10, 11 and 12. Later Mk 512 upratings were to add extra diaphragm discs.

2. Turbine entry temperatures, T4. Future increases were to lead to improved blade and disc materials and cooling techniques such as film cooling, and later to cast blades.

3. LP shaft torque. This was always a worry, which made uprating difficult as the LP shaft thickness was restricted by the HP shaft and bearings.

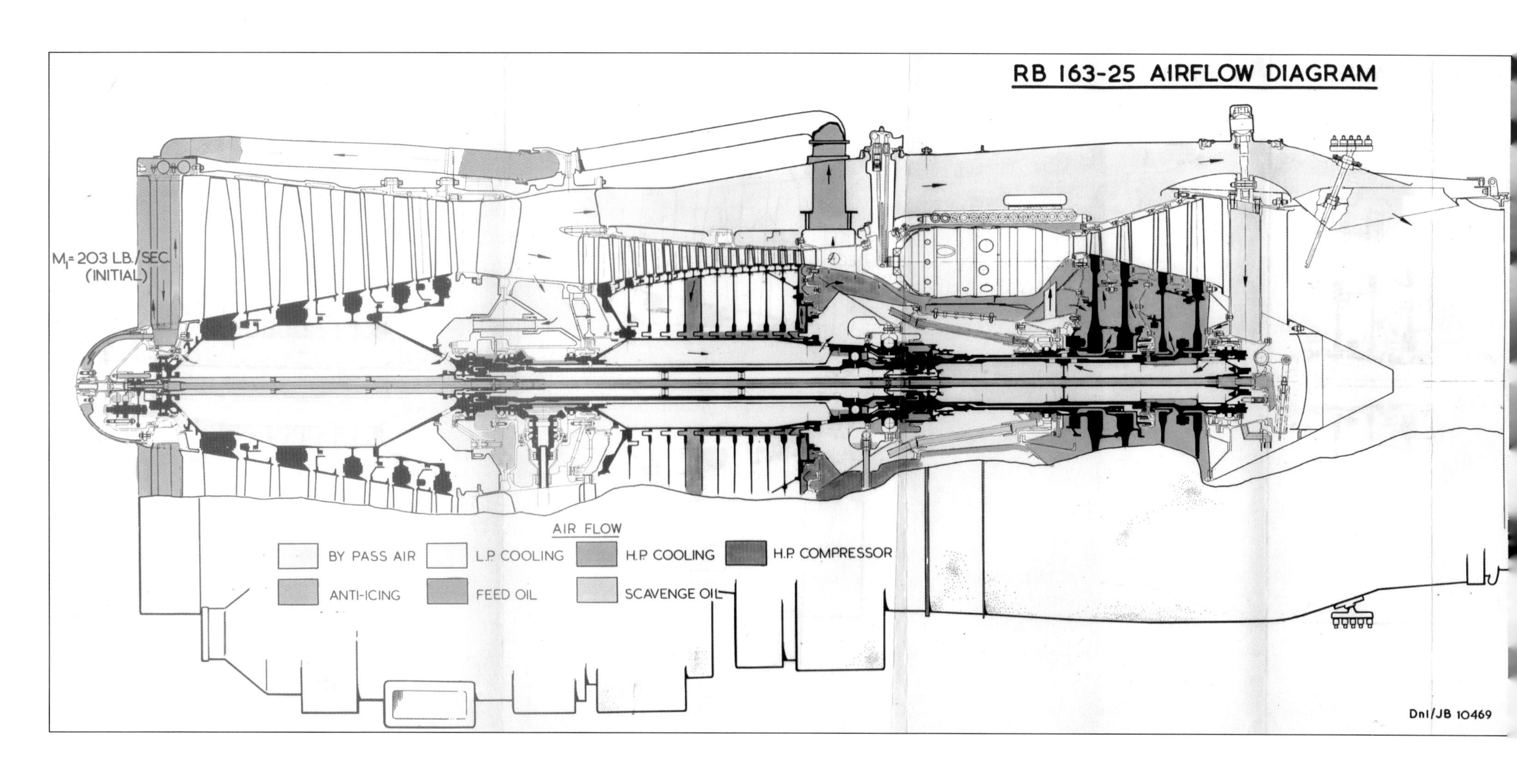

Figure 34
RB163-25 Airflow Diagram

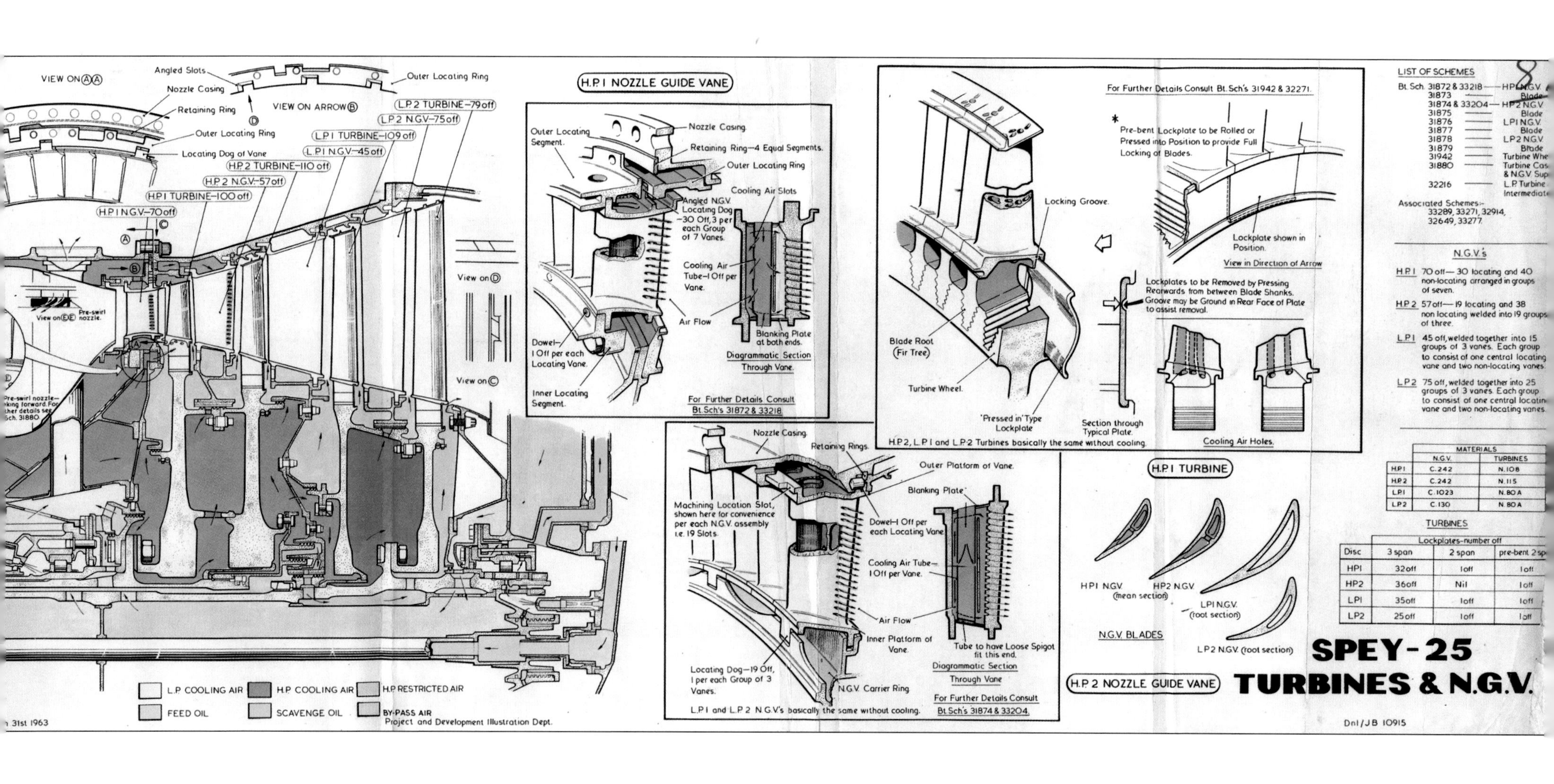

Figure 35
Spey - Turbines

Mk 510 and 511 (Trident 1E, BAC One-Eleven 3/400, Gulfstream II and III)

The 163-25 first entered service in 1965 in the Trident 1E. This had a 6 ft increased wingspan with larger fuel capacity to increase range and modifications to accommodate 115 passengers. A high-density version, the 1E-140 carried a maximum of 139 passengers – see Aircraft Chart in Appendix. Iraqi Airways had a 'Baghdad' rating of 2% higher wet takeoff thrust on very hot days – see Rating Chart in Appendix.

Figure 36
The enlarged Trident IE powered by the RB163-25

One year later the BAC One-Eleven-300 and –400 entered service. These were similar to the –200 but with strengthened landing gear and extra fuel capacity which, with more thrust, permitted an increase of passengers to 89 and longer range. The –400 was fitted with lift dumpers – spoilers at the rear of the inner wings to reduce landing distance. Surge was encountered during some reverse thrust landings with lift dumpers deployed at about 100 kts. Although it was not always detectable in airline operation, it did lead to HP compressor blade deflections of more than ¼" with resulting damage. Flight testing at Hucknall involving 200 landings, revealed that surging only happened with lift dumpers in operation, and was worse at higher NH and lower aircraft speeds. This led to revised operating procedures.

A most significant order for 30 aircraft came from American Airlines. At that time it was the biggest single dollar export order ever and established a very important commercial relationship which, apart from a hiatus when the airline chose the DC10 in preference to the Lockheed TriStar, continued with the order for the 535E4 in the B757, the Tay in the Fokker 100 and more recently the Trent 800 in the Boeing 777. AAL required a guarantee of parts cost, which was fixed at $7.21 per hour in 1963. Initial operation was at the 510 rating, when most expenses were due to the hot end, and this led to a programme of improvements to flame tubes, NGVs and turbine blades.

Figure 37
American Airlines became an important customer of the RB163-25 with the BAC One-Eleven -400

Grumman then provided a very successful home for the Mk 511 in their superior Gulfstream II, which followed the Dart-powered Gulfstream I. Cruising at 40-45,000' at Mn 0.72, it carried up to 19 passengers over 3250 nm range, and more with wing tip tanks. It entered service in 1967 and was succeeded 14 years later by the

G III, which had a new wing with more sweep back and winglets and a longer fuselage. It provided an 18% improvement in fuel efficiency, a large increase in range and higher Mn. In addition, Grumman produced a military version, the C-20A. It was used for maritime patrol and space shuttle orbital simulator work. Features of these models included a Rohr target thrust reverser and, to meet FAR 36 noise regulations, a hush kit of lobed silencer nozzle and a jet pipe with acoustic linings. To meet Grumman special requirements, the engine was tested with a fluctuating inlet pressure on the test bed and ATF.

Figure 38
The Gulfstream II powered by the RB163-25 (followed the Dart powered Gulfstream I)

Because the G II was a small aircraft with two large engines mounted on the rear fuselage, it was susceptible to high engine vibrations. A squeeze film bearing was introduced on the front LP roller, which had a different characteristic from the spring-mounted bearing. A gap of .002 to .005" between the outer race and the static parts permitted the bearing to take up an eccentric position. The gap was fed with oil, which was retained by two side plates locked to the outer race by dogs to prevent rotation. Back-to-back flight tests demonstrated a big improvement, and it proved to be a most effective modification.

In 1965 the Lear Liner 40 was launched in a most unconventional fashion. Bypassing normal sales contacts and procedures, Mr Bill Lear turned up unannounced at the commissionaire's desk at Moor Lane. He was seeking an engine for his new Model 40, a larger version of the 24 and 25, but smaller than the Gulfstream. This was for airline or corporate use, carrying 40 and 16 passengers respectively. In the latter role they sat in swivelling armchairs. After choosing the RB183-1 he switched to the more powerful Mk 511. A contract was signed, and Lear Jet and Rolls-Royce teams worked on modifications to the wheel case, accessories and pod. Giles Harvey recalls taking Danny Kaye, the American film star and comedian, around A Site in connection with this project; he was later Lombard's guest in the dining room where he entertained everyone (it is believed he was a Director of the Lear Jet Company and was concerned with sales). The project was later abandoned.

Mk 512 (Trident 2E and 3B)

The next civil uprating (about 5%) was the Mk 512 for two new Trident types for service starting in 1968. The 2E had a further 3 ft wingspan for more fuel to increase the range to 2400 nm. On the other hand, the 3B had a 16 ft longer fuselage to carry 31 more passengers. On top of this, there was to be a Super 3B for the People's Republic of China, with more fuel for increased passengers and range.

The 512 was rated at 11,950 lb takeoff thrust, flat rated up to 34°C with water injection. This was not enough for the 150,000 lb – 3B, which had to be fitted with a fourth engine (or fifth if you include the APU); this was the RB162-86 at 5250 lb thrust. The RB162 had been developed initially as a military vertical takeoff engine and as a consequence had a very good power-to-weight ratio. For the Trident –3B it was placed in the tail below the rudder, and was used as a booster for takeoff and climb out. The extra thrust could be used to increase the payload or reduce the takeoff run by 1800 feet. The booster inlet doors were closed when the booster engine was not in use to avoid windmilling drag.

Figure 39
Trident 3B on takeoff showing the installation of the RB162 booster engine above the centre RB163-25. The inlet doors of the booster close after takeoff

Half of the 5% extra takeoff thrust on the basic engine came from throttle opening corresponding to +20°C TET, and the other half from modifications to improve the thrust at a TET. Design features included:

- Long nose bullet.
- Front stages of compressor blades in aluminium had been prone to foreign object and other forms of damage. Several sets of titanium blades were made for trials, but were considered too expensive to introduce on existing Speys, compared with replacing damaged blades. A compromise on the 512 was for clappered Ti LP1 and 5 rotors, with a steel containment ring over the LP1, and also clappered HP1 rotors.
- -5° twisted LP1 rotors to improve high-speed efficiency.
- Turbine capacities adjusted to speed up the HP compressor by 380 rpm and increase airflow by 2½%.
- Two diaphragm discs between HP stages 10 to 12.
- Restored pressure loss flame tubes for improved hot end life.
- Upgraded materials and thicknesses in some compressor and turbine blades and discs, including N115 HP2 turbine blades.
- Slim line rear profile on the bypass duct.
- Six exhaust unit struts spaced as for 10.
- The two weakest points in the LP shaft were at the oil holes in the centre shaft, and the centre shaft and splines. Following a failure, the oil holes were smoothed and the surface layer compressed by pushing a ball through. To cope with the highest torque yet, the 512 shaft was 66% thicker than the –1, mainly on the less effective inner diameter. In addition, the material and heat treatment were changed, together with re-routing the cooling air to lower the temperature by 100°C. A fatigue test was satisfactorily completed on the 512.

Instead of operating the engine to full throttle at takeoff and checking the P7, the Mk 511 and 512 were set to a P7 corresponding to minimum certificated thrust +1%. On new engines this gave some alleviation in TET and other parameters which was beneficial to component lives without affecting safety. Of course, the benefit reduced as the TGT ran out of margin until overhaul when margins were restored.

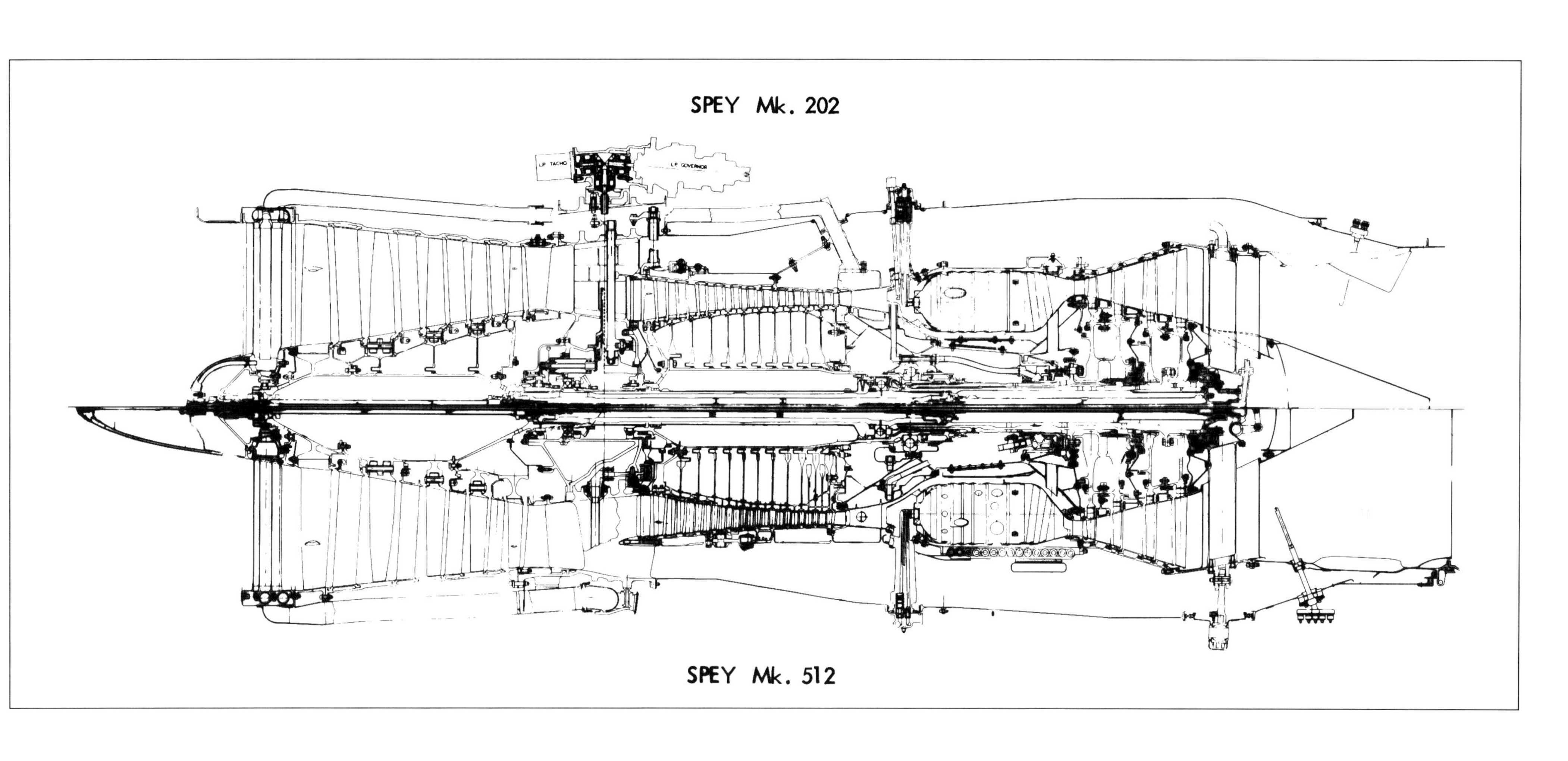

Figure 40
Spey Mk 202 & Spey Mk 512 Engine Comparison

Mk 512DW (BAC One-Eleven 475/500)

The next BAC One-Eleven development was the –500, with a 5 ft increased wingspan and 14 ft longer fuselage, to carry first 108 and then 119 passengers, with an increase in maximum AUW to 104,500 lb. It started life with the wet 512, but it was found that more takeoff thrust was needed at Berlin to avoid a tall railway station at the end of the runway. The answer was the Mk 512DW, which gave an ISA thrust of 12,550 lb wet only, flat rated to 25°C. This was achieved by raising the P3 limiter, and avoiding a small increase in NH by adjusting the –ve IGV stop on individual engines. Although this was not really a further throttle opening, it did mean that more takeoffs would be at the highest TET to the detriment of engine life. To compensate for this, a new method of setting up the engine on the runway had to be employed, known as flexible takeoff. Instead of setting to a P7 (P7-p0 or P7/P1) corresponding to minimum thrust depending on the runway, ambient temperature, etc, but regardless of aircraft all-up weight, it was set to a lower level when the aircraft was not fully laden. So that all takeoffs were not at the bare certificated safety level, a small margin was put in 'for the wife and kids' to make the pilots feel more comfortable; this system was the forerunner to that used today on all civil engines.

The BAC One-Eleven 500 went into service in 1968, and three years later came the 475 also with the 512DW. This had the larger wing span, but the original fuselage length and passengers. It was needed for shorter poorer grade runways. A 3% contingency rating was cleared for use following a critical engine failure during takeoff and climb-out, but was never followed up. The 512DW was destined to be the highest civil rating of the Spey. Production of BAC One-Elevens ceased at Weybridge in the 1980s, but continued in Romania under licence. The 500 and 475 became the Rombac 560 and 495. A corresponding programme covered manufacture of the 512DW in Romania. This is the subject of a Rolls-Royce Heritage Trust book by Ken Goddard.

In the mid 1970s, ICAO brought in obligatory noise standards for production aircraft. In a flurry of hush-kitting activity, all versions of the Spey sprouted exhaust suppressors and featured sound absorbent linings in the intake and tail-pipe to reduce compressor, turbine and combustion noise. On the 512 BAC One-Eleven, the tail-pipe lining necessitated an extended jet pipe with the nozzle sticking out at the back of the nacelle.

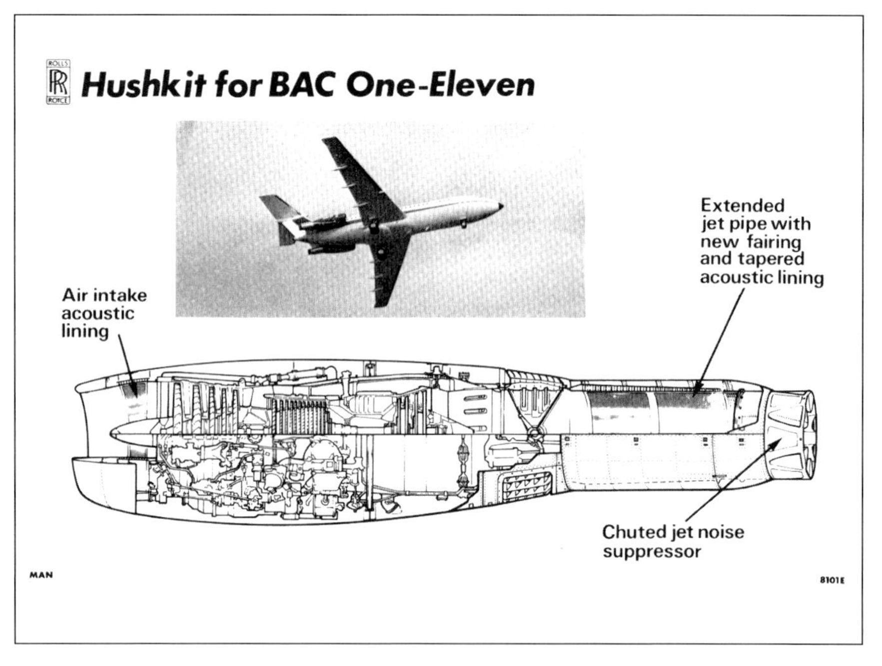

Figure 41

RB168-20 (Nimrod)

In 1964, design work was started on the Hawker Siddeley Nimrod to replace the Shackleton MR Mk 2 reconnaissance aircraft of RAF Strike Command. The HS801 Nimrod MR Mk 1 is a derivative of the Avon-powered Comet 4C, mainly used in a maritime reconnaissance role. This mission consists of a high-altitude subsonic flight to a search area, a low speed search just above sea level and a fast return to base. The Spey Mk 250 was suitable for this installation on account of its much better fuel consumption than the Avon, enabling a longer time at search.

Figure 42
The HS Nimrod maritime reconnaissance aircraft with two RB168-20 Spey engines being readied for installation

The installation layout is illustrated in diagram Dnl 14883. The intakes had to be enlarged for the Spey's higher airflow. The jet pipes of the inboard engines were too fat to go through the main wing spar and had to be made elliptical in the forward part. These pipes were angled at about 10º, and cracking was experienced in this section due to thermal gradients. Vortex generators were, therefore, fitted to throw hot air to the outside and promote more rapid mixing. The jet pipes were over 1½ times the engine length, more than enough to achieve exhaust mixing with an annular mixer rather than a chuted. Thrust reversers were fitted to the two outboard engines. The Mk 250 has the largest alternator of any Spey, and a modified accessory gearbox due to the very high electrical loads on the Nimrod.

The Mk 250 is aerodynamically similar to the civil Mk 511, with the takeoff thrust increased from 11,400 to 11,995 lb by throttle opening. Life was not impaired because of the lower utilisation of the higher ratings. Coming three years after the Mk 511, most engineering activity was associated with navalisation and the sea level patrol requirement. Materials had to be changed especially to prevent corrosion caused by prolonged ingestion of salt water, drawing on civil experience of the BAC One-Eleven with Aloha, who operated at low level between the Hawaiian Islands. Magnesium was largely excluded; titanium and stainless steel were favoured, and ferritic steels coated with resin-based paint to prevent oxidation. Immediately after return from a mission, the aircraft taxied through a pure water spray to get rid of the salt.

Much attention was given to obtaining the maximum time on the low level patrol where engines would be severely throttled back to obtain a mean speed of 200 knots. The SFC curve rises sharply at lower thrust levels due to cycle effects, lower component efficiencies at off design, and opening of the handling bleed valve. In addition, the SFC scatter between engines increases rapidly on this part of the loop to about 30%. By shutting down the two outboard engines and running on the inboards at double the thrust, fuel consumption is reduced very considerably. A limitation to this two-engine technique arises from the need to climb away in the event of failure of one of the inboard engines, particularly at the start of search before the fuel is burned off. Various options were considered for the outboard engines – idle, windmilling or intake blanking, and a full throttle contingency rating for hot days. The procedure adopted was to start the search above 200 kts with two inboard engines providing power, one outboard at idle, and the other outboard windmilling. After about three hours when the weight dropped sufficiently to enable a climb on one engine, the fourth engine was shut down to achieve the full fuel flow reduction. Eventually the speed could be dropped to 180 kts. In this way the search time with two engines is increased, for example to 6½ hours compared with 5½ with all four engines operating. In icing conditions, the windmilling engines were restarted and brought up to speed. Other possible problems arising from unusually long periods of running at idle or windmilling were compressor blade stresses at speeds below stall drop-out, and high oil consumption. Testing showed the engine to be satisfactory in these respects.

Mk 250 development testing started in 1966, the first flight was in 1967, and the Nimrod went into service in 1969. It has provision for six extra removable fuel tanks to be carried in the weapons bay, thereby increasing the maximum overload takeoff weight to 192,000 lb, the heaviest Spey-powered aircraft equal to more than 6½ AMXs. The Mk 250 is also in the Nimrod MR Mk 2 fitted with advanced communications and other equipment. The Mk 251 is a Mk 250 with uprated generator fitted in the Nimrod AEW Mk 3, an airborne early warning version used for detection and tracking in addition to the maritime role. The Mk 3 has scanners mounted in bulbous extensions of the nose and tail. It is still in service today but is due to be replaced by an updated version, the MRA4, featuring the BR700-710.

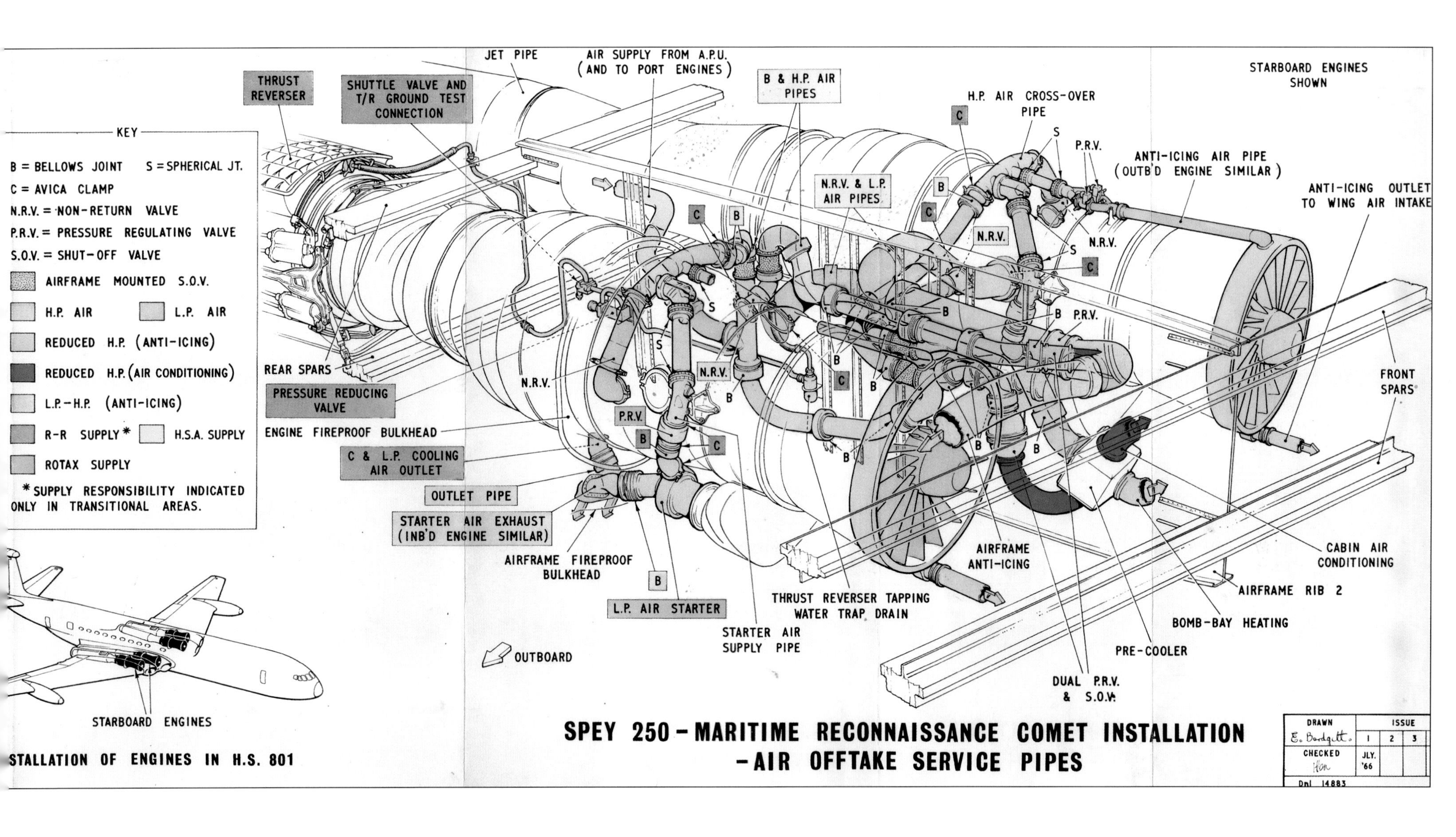

Figure 43
Spey 250 - Maritime Reconnaissance Comet Installation - Air Offtake Service Pipes

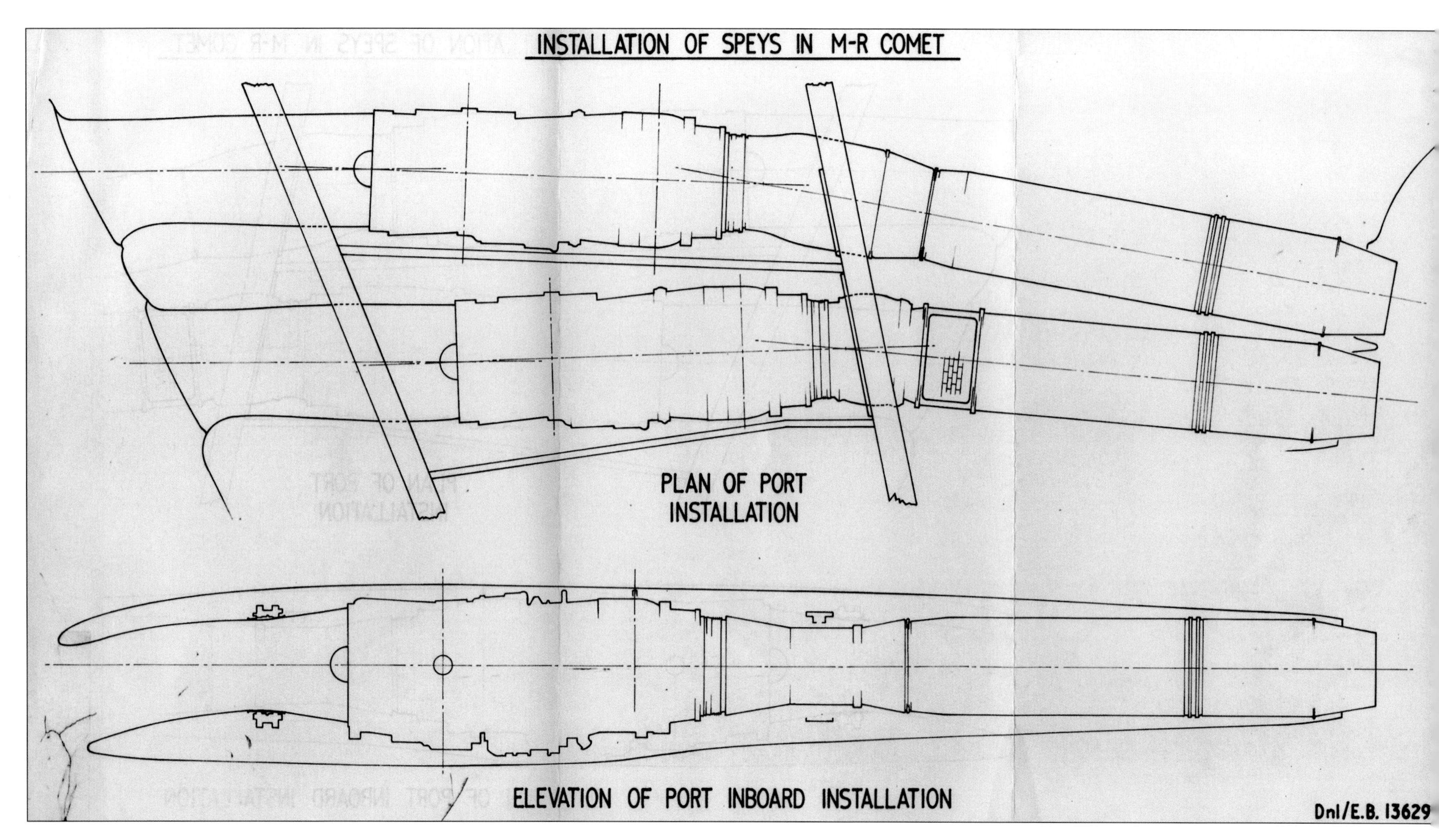

Figure 44
Installation of Speys in M-R Comet

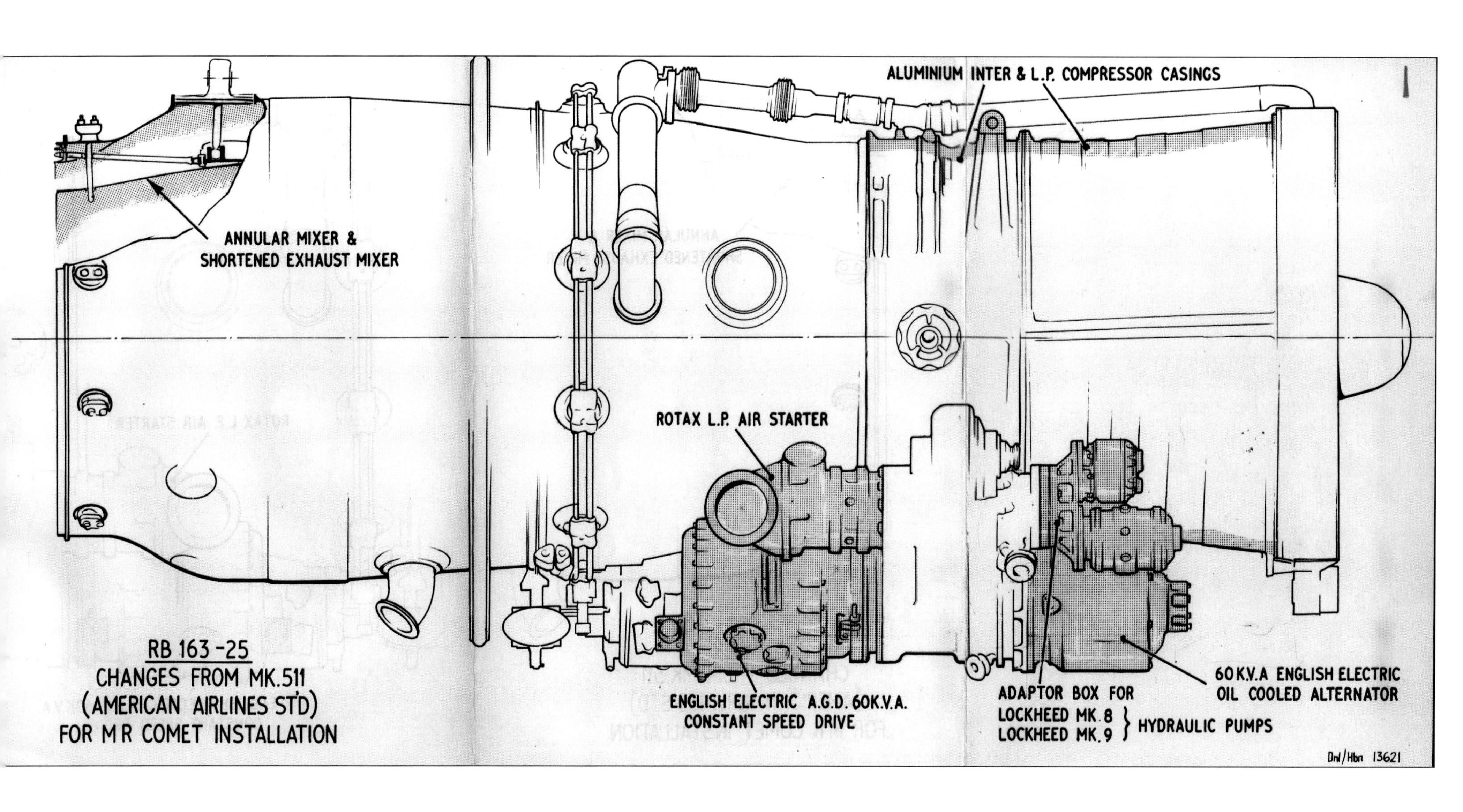

Figure 45
RB163-25

Mk 801 SF (Buffalo)

Pioneering work by de Havilland Canada on the augmentor wing concept, resulted in the conversion of a C-8A Buffalo transport aircraft into a research vehicle. The test programme was a joint venture between the Canadian Government and NASA, with Boeing and de Havilland Canada as the airframe and powerplant contractors. de Havilland subcontracted the engine and nacelle conversion to Rolls-Royce Canada.

The augmentor wing was a powered lift system for short takeoff and landing (STOL) at about 60 knots. This was achieved by bleeding off large quantities of engine air to be blown through slots in the wing trailing edges, and then deflected through two flaps. The high-energy ejector effect of the flaps induced secondary airflow over the wings, which gave a powerful increase in the lift coefficient.

Already in service, the Spey –25 was chosen for this project as it had the LP delivery airflow and pressure ratio (2.8) to meet the blowing requirements. Designated the Mk 801 SF, two engines were built with various Spey parts, but featured a new bypass duct to separate the hot and cold streams. The bypass air was transferred via two large ports on top of the engine to the wing, and the hot gas to a trouser piece nozzle system of the Pegasus type. The hot nozzles could be swivelled between 18½° to 116° below horizontal in order to keep the aircraft on the required glide path without having to reduce engine power and hence wing lift.

The 801 SF project was started in mid 1969. With Rolls-Royce Canada in overall charge, design, manufacture and build was carried out at various Rolls-Royce sites, including Rolls-Royce Canada. Testing commenced at Montreal in 1971. Two major actions were necessary:-

1. A pressure-reducing colander plate with about 400 holes 1" diameter had to be fitted downstream of the LP turbine to reduce turbine vibration. The colander also reduced jet noise and thrust, but the latter was more than adequate anyway.
2. Unlike other Speys, bleed valves had to be fitted on the bypass duct to obtain satisfactory LP compressor handling, especially during deceleration. Running with separate jets changed the slope of the LP working line. More importantly, the flow characteristics of the wing slot system affected engine speed matching, and the larger exit volume resulted in different transient working lines due to the packing effect. With the Avon-type bleed valves, surge-free handling was achieved.

The initial flight test programme lasted from 1972 to the mid 1970s, concentrating on aircraft handling, steep descents and automatic landing. Noise was not important except on steep landing approach, so a silencing package consisting of an acoustically-lined front bypass duct and nose bullet was fitted to the engines. The Speys were updated and cleared for further flight testing into the 1980s making a grand total test time of 1000 hours.

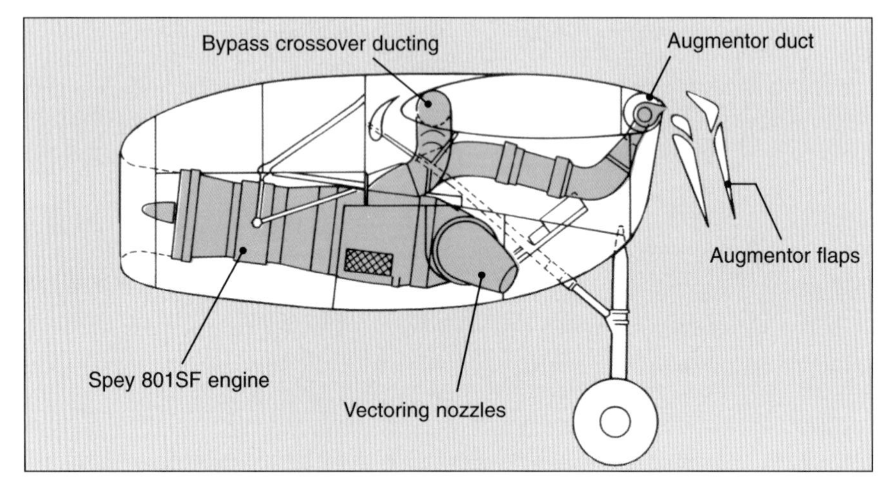

Figure 46
The installation arrangement of the Spey 801st engine in the Buffalo

Figure 47
The Buffalo Aircraft Incorporating the Four Augumented Wing for Short Takeoff and Landing

Chapter six: RB168-25R – Phantom Spey

Figure 48
McDonnell F4M Phantoms powered by two RB168-25R Speys

In 1964 the UK Government decided to buy the McDonnell Phantom fighter for the Royal Navy (F-4K); later this order was increased to provide aircraft for the Royal Air Force (F-4M). This was a proven aircraft and there was no British equivalent. However, there was a British engine – the Spey – that could be adapted to fit the Phantom, giving advantages of enhanced performance over the General Electric J79-8 straight jet used by the American forces. In particular, it could offer the thrust needed to allow the aircraft to takeoff using the aircraft carrier Eagle's waist catapult, which was much shorter than US catapults.

Although the RB168-25R was basically a civil –25 engine with reheat attached, it merits a separate chapter on account of its very great importance to the Company, the severe requirements, and its technical complexities and difficulty compared with other Speys. It was the first reheated bypass engine to fly behind a mach 2 supersonic intake and it did it very successfully in America's backyard, in contrast to the PW TF30, which suffered considerable handling problems in the F111.

Requirements of the engine included:

1. Fully variable 70% boost reheat for takeoff, bolter[2] and acceleration to Mn 2.0, 36,000'.
2. Carrier takeoff in reheat from 150' and 250' catapults (US Navy 350').
3. Boundary layer control (BLC) air from two different HP compressor stages, 7th and 12th; 7th for takeoff and 12th for landing. Effectively, the control system had to cope with six different engines, zero, 7th and 12th stage BLC, reheat lit and unlit.
4. Designed for Mn 2.2 at 36,000', severe inlet pressure distortion, manoeuvres, cyclic duty and maritime conditions.

[2] Bolter is a term used when an arrester wire is either missed or breaks during a carrier landing and there is an immediate need for full reheated takeoff thrust to become airborne again.

5. Variable intake, spill valve and translating shroud ejector nozzle.
6. In addition, it nearly had a Moog valve – in fact, it had a full set of 'bells and whistles' except for water injection, a bog chain (LP shaft failure protection device) and thrust reverser.

The Spey met these requirements and, being a bypass engine, offered higher takeoff and subsonic thrusts, especially in reheat, and better cruise SFC than the J79. These were in addition to the advantages of reducing the dollar cost by about 50% as the engine and other airframe parts were paid in sterling. The choice also gave Rolls-Royce the chance to develop its engine technology, particularly reheat.

The engine specification was extremely detailed and rigid. Rolls-Royce had to meet 26 performance guarantees as well as those concerning engine dimensions, interfaces, time at all stages of the programme, etc. Although the engines were sold to the British MoD, all clearance work for the F4K for the Royal Navy was to be done by the USN and BUWEPS. On the F4M, Boscombe Down was involved for weapon clearance, etc, for the RAF.

Engine design

The –25R engine was aerodynamically similar to the civil Mk 512, with the throttle opened to give 12,250 lb reheat unlit takeoff thrust at the increased TET of 1425°K, boosted to 20,515 lb at full reheat.

It was designed at the same time as the Mk 512, so there were many common features. Differences arose due to much higher temperatures and pressures from LP compressor inlet to HP outlet and for other operational reasons, such as the need to cope with severe inlet pressure distortions due to operating behind a Mach 2 plus supersonic inlet and the higher manoeuvre loads of a carrier-based fighter aircraft.

Features of the –25R were:

- Clappered LPC1 blade in titanium to cope with bird and ice ingestion (intake lip not anti-iced).
- Clappered LPC5 blade to prevent choked flutter failure if the final nozzle opened with reheat unlit.
- LPC shaft and discs instead of drum to strengthen.
- LPC casing changed from Mg to Al, for anti-corrosion at sea.
- Steel intermediate casing.
- BPD reduced in diameter by 0.7" to fit in the tight installation.
- HPC1 clappered blade to cope with ingestion.
- HPC blades in Ti and Ni alloy.
- HPC stators in steel and Ni.
- Rear HPC discs in Ni instead of steel, for high temperature and lightness.
- HP compressor 1¼" longer, larger gaps to avoid blade clipping following surge, a recurring problem on the civil engines but more severe due to higher pressure levels on a military aircraft.
- Stronger engine mounts for high axial accelerations during catapult launchings and arrester-hook landings.

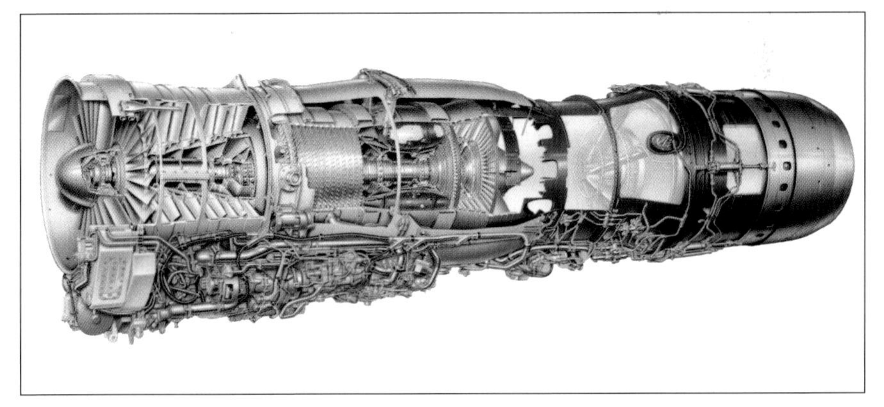

Figure 49
A cutaway drawing of the RB168-25R showing the engine reheat systems and nozzle. The accessories are mounted under the engine

By the time these design changes had been made, plus further modifications during development, there were few parts left which were common to the civil engine.

The Phantom required BLC air for wing leading and trailing edge flaps during takeoff and landing. For carrier approach with full flaps and the engine throttled back with no reheat, the high-pressure air required for BLC was supplied from the 12th stage (HP compressor delivery), normally 14.3 lb/sec. For hot day takeoff at full engine throttle, reheat lit, half flaps, the temperature and pressure of 12th stage air were too high, so 7th stage was used, normally 6¼ lb/sec. During a landing, the engine was slammed from approach thrust to full throttle at the point of touch down to ensure a successful bolter could be achieved. If the pilot waited until he was aware that the hook had missed the last arrester wire (there were 4 of them) it would already be too late and the aircraft would sink into the sea as it left the deck; thus the switching system between 12th stage bleed and 7th had to be very rapid to avoid over-temperaturing and over-pressurising the aircraft ducts.

Reheat background and design

The –25R was required to give at sea level static, reheat boosts from 7% to 68% at full engine rating (NH, T4 etc); 68% boost corresponds to a mean reheat temperature of 1950°K, and it was obvious that this could not be achieved by burning in the hot stream alone. Previous service experience had been on the straight jet Avon burning in a 950°K turbine exhaust. The considerably lower mean exhaust temperature of a bypass engine required a higher temperature rise, and the lower jet pipe pressure did not help burning. The hot exhaust temperature of the –25R was 950°K, but burning in the 2" wide cold stream of 410°K presented a problem, so the decision to mix the streams was taken.

Engine reheat test programmes were carried out on the RB141 in 1961 and the RB153-61 in 1963 – on mix, diffuse and burn reheat systems (see below for details). Both of these engines had cycles and jet pipe conditions similar to the –25R, so it was possible in 1964 to base the –25R on this experience. The aim of the testing was to get efficient and stable burning in the mixed exhaust streams between 6% and 70% boost. Traditional vee gutters were unable to keep at the nose of the stability loop on a bypass engine over a range of boost. This was solved by having fuel-fed vapour gutters scheduled with pressure to keep a constant fuel/air ratio (90% of FAR at the nose of the loop) at all conditions. The width of the loop was extended by fill fuel, which was staged with degree of reheat.

During a US competition to select an engine for the new swing-wing F111 (the TFX project) aircraft, the RB141 was fitted with a 44¼" reheat pipe for testing at Indianapolis; see *Fast Jets* by Cyril Elliott. Testing involved different mixers, diffuser and pipe lengths, gutter configurations and inter-connectors. High boosts were obtained and satisfactory modulation. Altitude tests were run on a model. The reheated bypass TF30 (military version of the JT8D) was eventually selected for the TFX.

Testing on the smaller MAN/R-R RB153-61 was directed at the VJ101D, and funded by the German Ministry until it was cancelled. Fitted with a 32" pipe, the programme included work on fuel distribution, reheat control, variable convergent-divergent nozzle, ignition and reheat accessories. In 1964 the RB168-1 was tested with a scaled RB153-61 pipe.

The –25R reheat pipe was basically the RB153-61 scaled from 32" to 37½", somewhat less than the 41½" which would have given the same pipe Mach numbers. The design incorporated:

- A 10-chute mixer (later 20 chutes) and fatter exhaust cone.
- Burners after the diffuser; gas velocity approx 500 ft/sec.
- Flame stabilisers, three vapour gutters with inter-connectors, separately fed with fuel at constant FAR for best efficiency and stability.
- Main (fill) fuel for boost from four manifolds, with staging according to degree of reheat.
- Ignition by fuel-fed catalyst (replacing hot shot), ceramic with platinum-iridium plug for fast light up.

- Outer wall heat shield taking 10% of air, some re-entering to burn, the rest cooling the nozzle.
- Variable area primary nozzle, consisting of a number of overlapping flaps hinged to a ring. Nozzle position was controlled by rollers on an actuating ring, which acted on a curved profile on the back of the flaps. The ring was part of an axial sleeve positioned by six hydraulic rams.
- A fixed geometry secondary nozzle was attached to the actuating sleeve. At the reheat Vmax condition, this provided the correct con-di area ratio. For reheat unlit, spilled air from the supersonic intake ventilated the gap between primary and secondary nozzles, thus minimising the base drag.

Control system for engine, BLC and reheat

The control system for the main engine was the same as for the civil engines, ie, CASC, governors and limiters, NH, NL, P3, T6, except for:-

- The combined HP IGV and handling bleed ram and throttle cambox of the civil Speys, which took its N^2 signal from the HP governor, was replaced by an $N/\sqrt{T_1}$ governor which generated its own N^2 signal by internal fly-weights and took the T_1 signal from the amplifier, a separate IGV and BV ram and a separate cambox which contained all the reheat command functions as well as the main engine.
- T3 limiter to prevent over-temperaturing of the combustion casing.
- NH and T6 limits varied with T1 (but see later in Flight Testing).
- T6 re-datum to give +4% thrust for hot day carrier takeoff, reheat lit, 7th BLC (C rating).
- Fuel 'Big' Dipper, to recover from deep compressor stall (also added to civil engines).
- 7th and 12th stage BLC controls, sensing, switch-over and resets to ACU, FT and PRCU. A switch-over valve linked to half or full wing flaps selected 7th or 12th stage air. The quantity was fixed by the LE and TE flap slot areas, but choked venturis in the engine pipework prevented excessive bleed if an engine or aircraft duct failed.

A new system of reheat control had to be devised. On previous engines, the pilot's throttle lever selected nozzle area and pressure signals chose the reheat fuel flow to restore the engine matching. This would not work on the –25R because, at maximum reheat, fuel was burning almost stoichiometrically, so changing the fuel flow had negligible effect on pressures. Therefore, the system had to be reversed. In any case, the control parameters needed to be fundamentally different because of the different matching of a bypass engine.

A maximum reheat fuel flow (FR) scheduled against P3 would give approximately constant FAR and, therefore, constant reheat temperature rise. However, the need to restrict FR at one condition due to buzz could compromise the reheat thrust at another condition with the same P_3. For example, P_3 at sea level takeoff is almost the same as at 36,000 ft Mn 2.0, even though the non-dimensional conditions are different. The solution was to bring in another degree of freedom by generating a split P_3 (P_3') using a non-dimensional P_3/P_2. Fuel was delivered by a Dowty vapour core reheat fuel pump.

When selecting the engine pressures for final nozzle control, it is important that for a given change in nozzle area there should be a large shift in the pressures relative to the shift in LP compressor working line. This ensures that following an increase in FR the nozzle moves swiftly and decisively to prevent LP compressor surge. By controlling the nozzle on a P_3/P_6 v P_3/P_2 schedule, it is possible to restore the LP working line to the unlit level. There are advantages at high mach number to open the nozzle further to increase the engine airflow, because this wins out over the lower exhaust pressure and increases the thrust. This is called 'under restoration' and was acceptable as it also increases LP surge marin; it also suited the intake airflow matching and reduced the need for spillage at high mach number.

To achieve a fast and precise actuation system, the nozzle rams were hydraulically-operated rather than pneumatically as in earlier Rolls-Royce engines – they were also smaller easing installation problems. High-pressure oil, controlled by engine pressure signals was supplied from a separate engine-mounted pump. A separate oil tank and cooler was also used to avoid the aeration present in engine oil.

The reheat lighting sequence from maximum dry rating was a +10% nozzle pre-open to prevent LP compressor surge due the sudden rise in pressure when the fuel ignited. Fuel was fed to the vapour gutters and ignited by the catalyst. The nozzle then became live and was controlled by the pressure ratio control unit. With the vapour gutter fuel constant, increased boost was obtained by increasing main fuel injection, the distribution to the manifolds being controlled by a staging valve; initially there was a feedback to the staging valve from the nozzle to control the rate of acceleration but this was deleted before first flight as it caused surging on accelerating to maximum reheat on takeoff. To prevent surge on reheat acceleration, the rate of increase of the reheat fuel was controlled by a spill valve in the fill fuel line, which was gradually closed. A smooth transition from reheat unlit to maximum reheat was eventually achieved. Quick light up for a baulked landing (bolter) was helped by reducing the volume of the fuel feed pipes to a minimum.

Accessories and installation

Although the –25R was the same length as the J79 (17 feet), its airflow was 20% higher. To accommodate the fatter engine, McDonnell swelled the airframe and inlet duct, and Rolls-Royce reduced the diameter of the bypass duct. Even so, it was an extremely tight fit with clearances in general at 0.5" and 0.050" in some places; it was possible to look into the engine bay with no engine in it and see a mirror of the external shape of the engine – rather like a mould. Each engine had to be passed off through a very complex series of templates.

Figure 50

An RB168-25R on a trolley prior to installation in a RN F4 Phantom at the McDonnell plant in St Louis. The trolley shown is a US Airlog trolley in RN service, a specially designed R-R installation and transport trolley was used

Service access to the engine was from underneath, so accessories had to be mounted below the engine. In addition to the normal Spey engine accessories such as fuel control unit and pumps, engine oil pumps and cooler, aircraft hydraulic pump, tachometer generators, etc, there were additional units associated with reheat and a computer as big as a shoe box in a fuel-cooled casing, which was part of the control system. Also the starter and constant speed electrical generator were double the civil capacities. To fit all this in a small space without over-stressing the mountings, two gearboxes were fitted. The normal one on the intermediate casing drove engine accessories, and a new auxiliary one below the intake casing was for aircraft use.

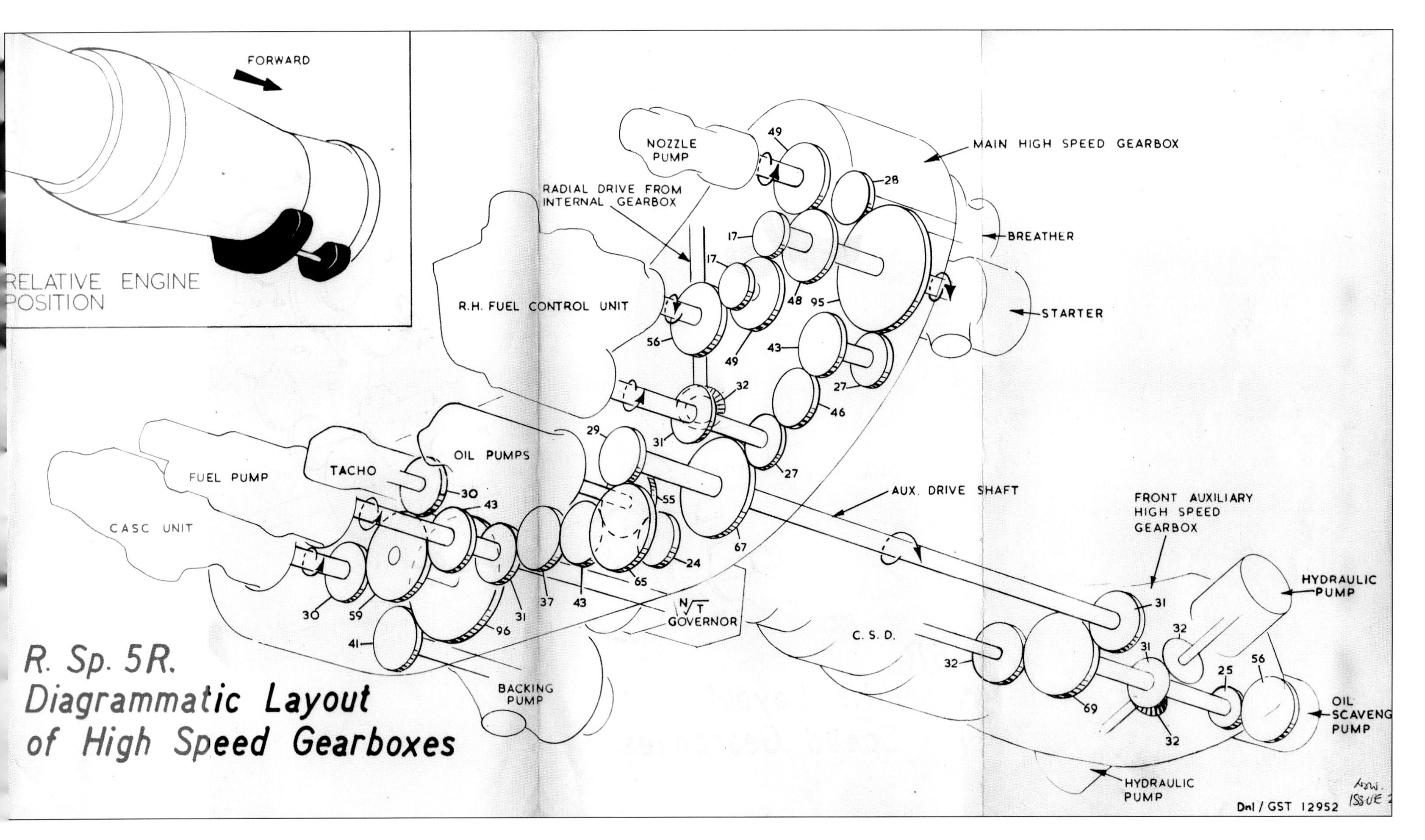

Figure 51
R Sp 5R Diagrammatic Layout of High Speed Gearboxes

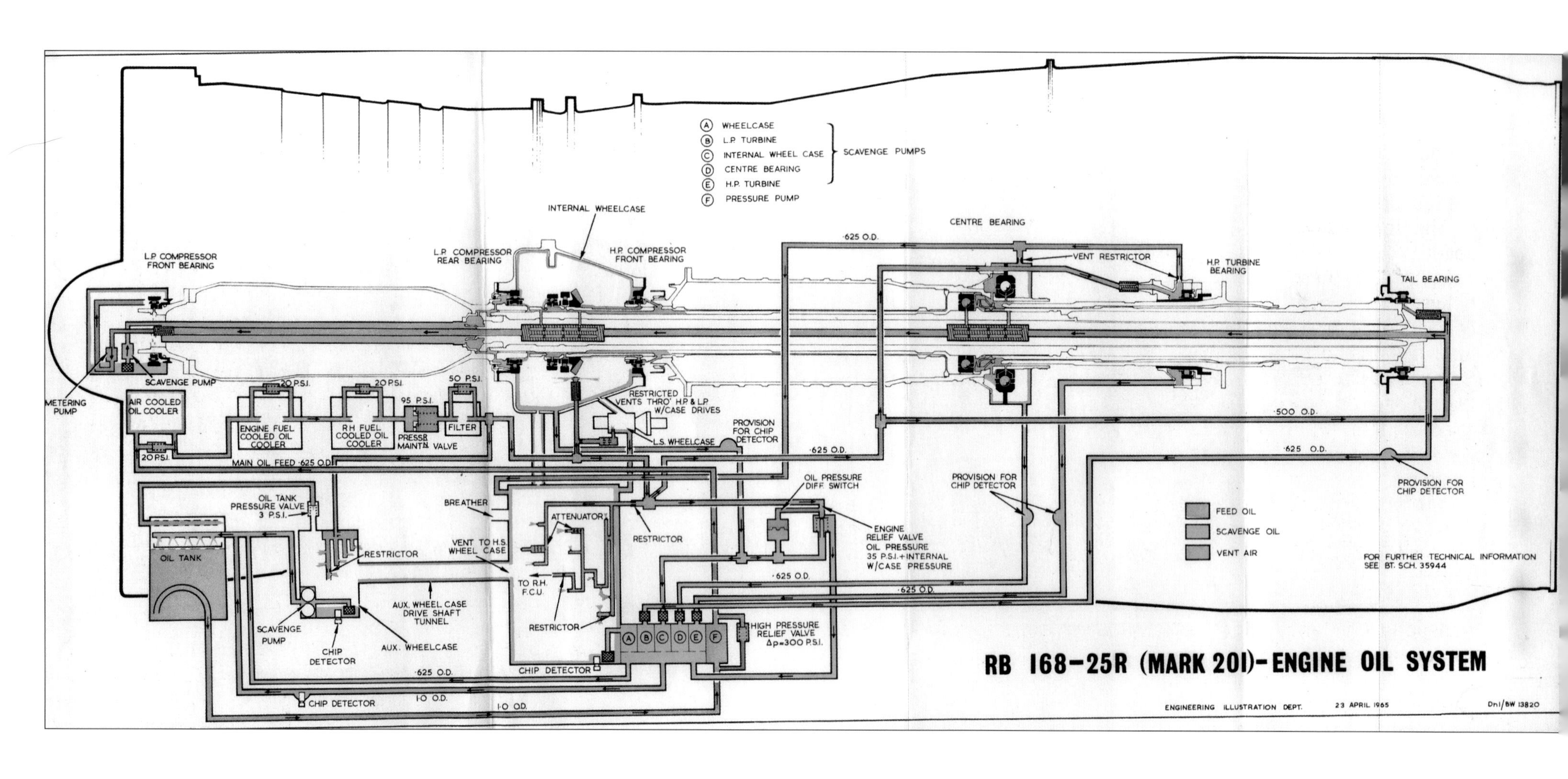

Figure 52
RB168-25R (Mark 201) - Engine Oil System

Unlike civil installations where air offtakes were required in the stub wing opposite the HP compressor, the Phantom anti-icing and BLC air were needed in the plane of the turbines. All these services had to be piped down the inside of the BPD in the so-called 'bagpipes'. Outside the engine were 289 pipes lying 4 deep and, as Les Buckler once laconically remarked *'all different'*.

Oil cooling was a real dog's dinner as there were three separate oil systems (engine, constant speed drive and reheat nozzle), each with its own tank, filler and cooling system. In addition, two of them needed dual coolers. Fuel cooling was employed in cases where air was hotter than the oil; in other cases air was used to prevent the fuel boiling.

The specification required that the starting equipment should be self-contained within the aircraft. To achieve a fast start each engine had to be started simultaneously, which ruled out cross-feed between engines. The outcome was a miniature gas turbine consisting of a centrifugal compressor, annular combustor, one turbine stage to drive the compressor and a second to start the Spey. The starter was started by a small electrical motor no bigger than a teacup, and powered by the aircraft battery. The MoD specified the starter should be supplied by Plessey. It was not available until the 50th engine, so the engine began service with an air starter.

The bottom access door was too small for installing the engine. The J79 could be wheeled in through the back end. The Spey was too fat for this, so it had to be manoeuvred in along a not-quite-straight path. This required the design of a special installation stand, which was built up by adding components to the basic transportation stand; it also had to work with extreme precision on a carrier in a rolling sea!

There were two engine mounting plates, one on the compressor intermediate casing, and the other at the rear of the jet pipe diffuser. The reheat pipe was cantilevered rearwards, and the diffuser casing was made of thicker material. Fred Steele of McDonnell writes:-

> *'I often pointed the Spey mounting out to anyone interested in such things, and bragged about how much better it was compared with the .... All the thrust was taken at one inclined ball joint, and there were ball joints at the other two points where the joints met the airframe. All loads were determinable, and the engine could expand or contract as it chose, and the airframe could bend or deflect in any way it needed and not burst the engine.'*

The intake, which was a high mach design with variable ramp to control the shock position, was scaled up to handle the 20% higher airflow at takeoff. However, at higher inlet temperatures, the non-dimensional flow falls off more rapidly on a bypass engine than a straight jet. The variable intake was scheduled against inlet temperature, and to utilise the same control the Spey had to be steered between narrow airflow limits. Any differences in intake airflow and engine demand were accommodated by a circumferential spill valve fitted just fore of the engine inlet. At takeoff the spill valve was used as a supplementary inlet as air was sucked from the engine bay to augment the flow from the main inlet. As forward speed increased, flow through the valve reversed and secondary air passed round the engine to the final nozzle, cooling and ventilating the engine bay and reheat pipe and by ejecting between the shroud and nozzle, reducing base drag at subsonic reheat unlit conditions.

Development testing

Considering the amount of change required to the basic civil engine and the degree of technical innovation, the whole design and development programme was extremely tight. This demanded everything be more or less right first time which, in turn, required very thoughtful design and Herculean efforts in development and manufacturing. Initiation of the programme was 01 July 1964, first engine run was mid 1965, and the first flight engines were delivered in the spring of 1966 with first flight on 27 June; less than two years from initiation of the programme to first flight for a virtually brand new, amazingly complex, ground-breaking technology, military engine!

As might be expected, most performance and mechanical problems encountered during the development programme were

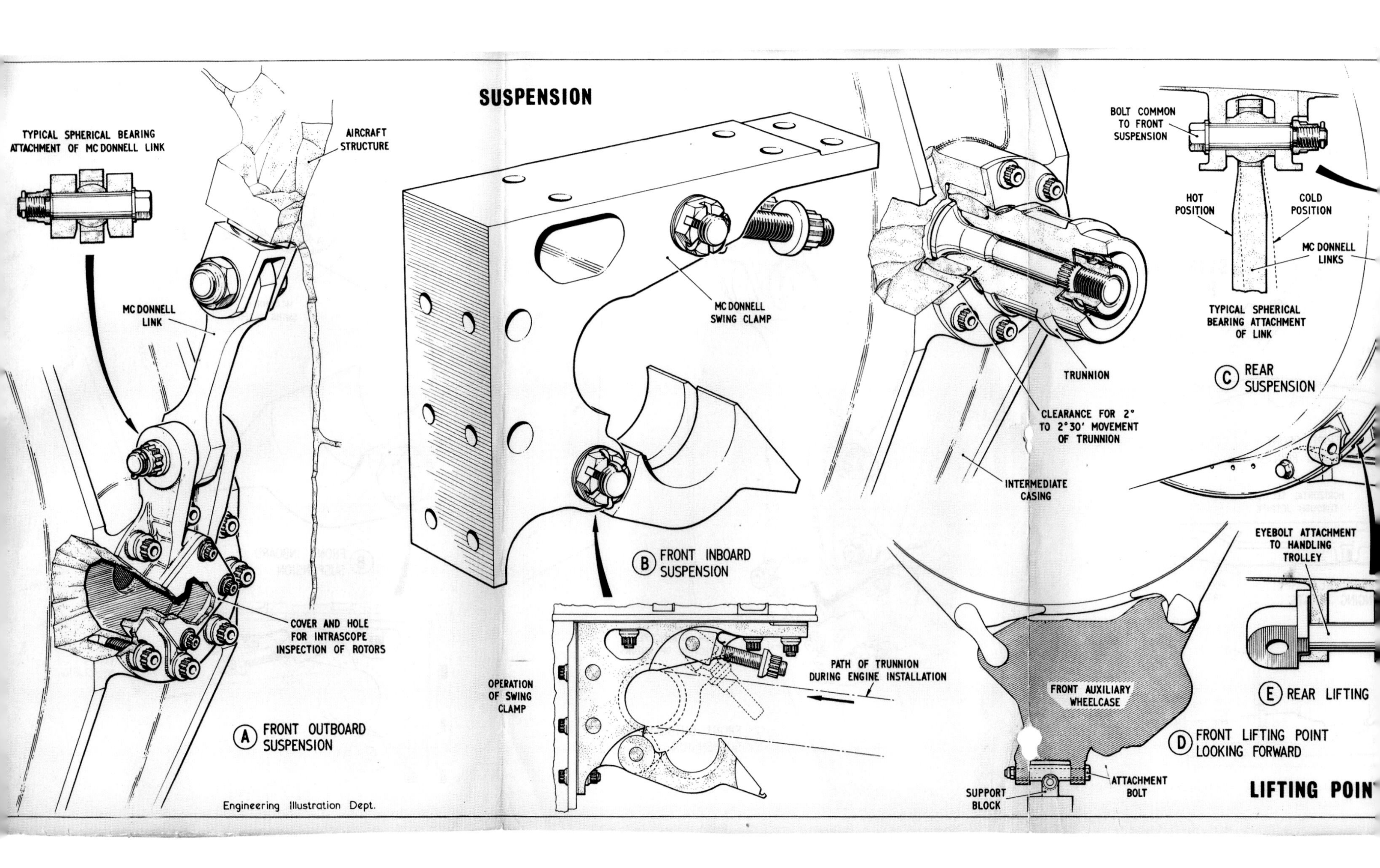
SUSPENSION
TYPICAL SPHERICAL BEARING ATTACHMENT OF MC DONNELL LINK
AIRCRAFT STRUCTURE
MC DONNELL LINK
COVER AND HOLE FOR INTRASCOPE INSPECTION OF ROTORS
A FRONT OUTBOARD SUSPENSION
Engineering Illustration Dept.
MC DONNELL SWING CLAMP
B FRONT INBOARD SUSPENSION
OPERATION OF SWING CLAMP
PATH OF TRUNNION DURING ENGINE INSTALLATION
TRUNNION
CLEARANCE FOR 2° TO 2°30' MOVEMENT OF TRUNNION
INTERMEDIATE CASING
BOLT COMMON TO FRONT SUSPENSION
HOT POSITION
COLD POSITION
MC DONNELL LINKS
TYPICAL SPHERICAL BEARING ATTACHMENT OF LINK
C REAR SUSPENSION
EYEBOLT ATTACHMENT TO HANDLING TROLLEY
E REAR LIFTING
FRONT AUXILIARY WHEELCASE
D FRONT LIFTING POINT LOOKING FORWARD
SUPPORT BLOCK
ATTACHMENT BOLT
LIFTING POIN

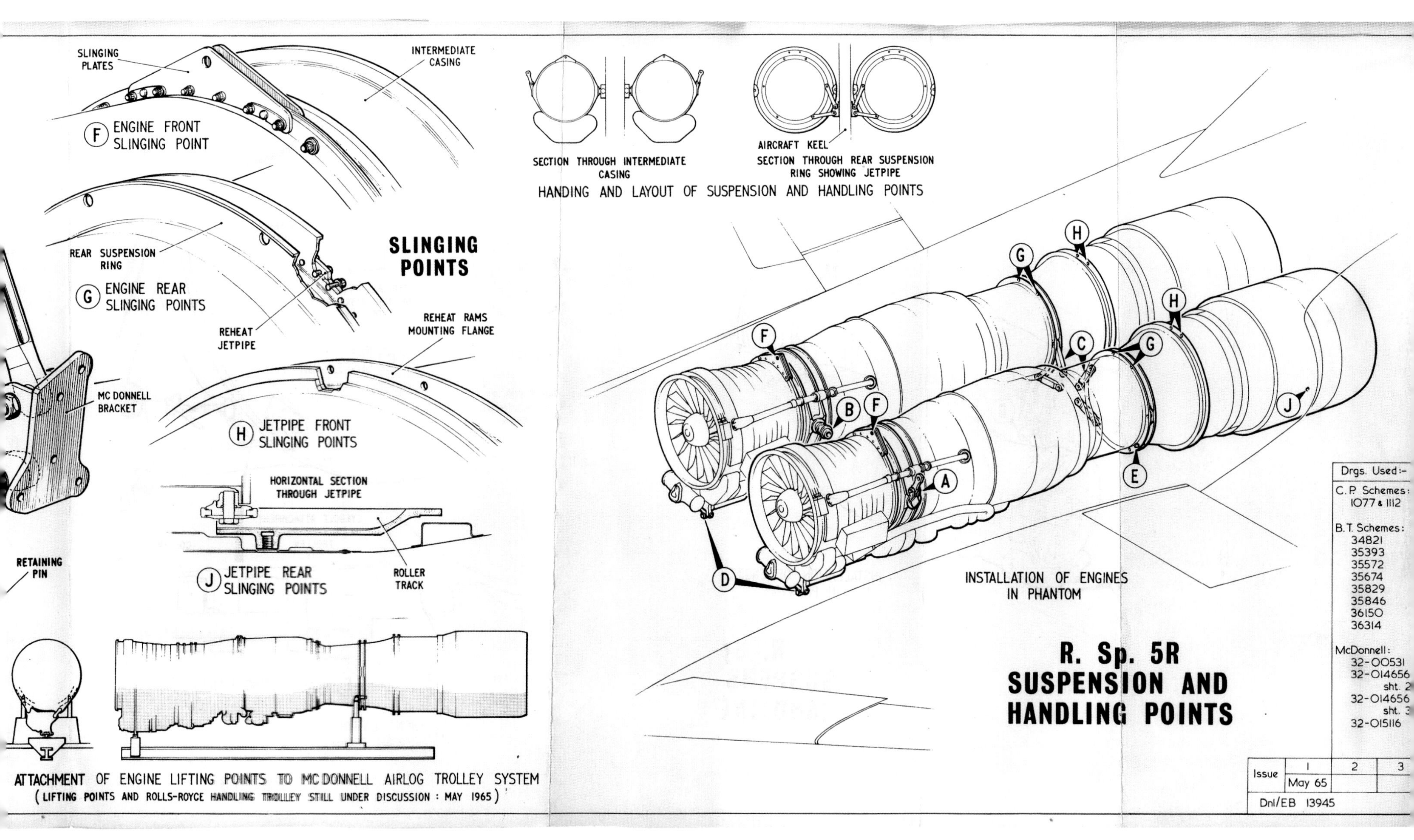

Figure 53
R Sp 5R Suspension and Handling Points

associated with features not common with the civil equivalent. A very important exception was severe overheat due to HP compressor stall, which afflicted all Speys and other types of engine – the Adour, at a later date, for example. The condition could be triggered in a number of ways, but the result was engine stagnation with a rapidly rising temperature and burn-out of HP turbine blades. The cure was to fit a dipper, which rapidly chopped the fuel to idle. This rather technical story is explained in more detail in the appendix.

As on the Buccaneer, there were failures of the handling bleed valve. The ring was moved axially by one ram and, in order to cure jamming, the clearance between the ring and the two lands was increased. This reduced the damping of the mechanism, resulting in vibration of the ring in a swash mode, and fatigue failures in the region of the operating lug. The problem was cured by adding other operating lugs to the ring with associated external linkages to the compressor casing; this was known as the 'double drive bleed valve'.

As soon as the engine was operated in the altitude plant at Mn 2.0 inlet conditions, the intershaft bearing between the HP and LP shafts failed releasing serious amounts of swarf into the oil system. Measurement of the end load on the LP shaft showed that at this condition the failed bearing was operating at zero load and skidding. This attacked the balls, reducing them uniformly in size. After several attempts to move the zero point, a special additional back-to-back bearing was installed. This bearing had a separating load applied by a Belleville washer, which was large enough to prevent the load on the locating bearing from reaching zero. After achieving the correct operating load for the Belleville washer, the problem was solved.

The –25R was first run without BLC pipework and reheat pipe, and gave performance equivalent to the civil level. When these features were incorporated, the SFC worsened much more than expected, especially when the HP compressor working line was lowered to ensure good handling. One of the ways out of the reheat thrust problem, which was to follow, was to improve the main engine performance and thus lower the reheat temperature. In addition, there was a need to get more TET margin and simplify the many operating limits. McCarthy was wheeled out again with lists of sealing, blanking, fairing, IGV optimising and rematching modifications. Gradually these were to improve the performance, but more effort was needed to achieve the reheat target.

Buzz limited the maximum reheat boost; it could be heard and felt, and ultimately damaged the reheat pipe. It was caused by the effect of the non-uniform flow from the mixer, and resulted in cyclic blow-out and exploding re-ignition of the fill fuel flame fronts, ie, 'buzz'. This instability limited the thrust at high altitudes, and also restricted the reheat burning and lighting altitudes. Initial ATF testing revealed reheat thrust deficiencies at 36,000' up to 14% because of under-fuelling to avoid buzz. This gave rise to the blue engine, which was an interim standard operating without buzz, but with some thrust deficiency. The red standard followed later with modifications to recover the thrust, especially at the all-important 36,000' Mn 2.0 case.

The blue standard mainly featured an FR schedule trimmed to avoid buzz, but also had a modified fuel distribution, wider gutters and a 20-chute mixer. The standard expected at 36,000' was a deficiency of 10% thrust at Mn 2.0 and somewhat less at lower speeds. Blue engines easily met this level.

The red standard had:-

1. A kink in the PRCU schedule, which raised the engine airflow at high Mach numbers.
2. A modified FR schedule.
3. Change to vapour gutter fuel distribution.
4. Blanked No 1 fill fuel manifold.
5. Engine set to test bed thrusts above 12,250 lb unlit.

A number of engines achieved the specification thrust at Mn 2.0, but with the arrival of better component efficiencies in the engine the buzz level deteriorated. The reason for this paradox was that as performance improved NH rose and, because of the NH limiter, effectively throttled back the engine at Mach 2.0 to well below

Figure 54
A spectacular shot of the RB168-25R on test at full reheat. Of interest are the multiple instrumentation lead-outs and the ingestion of the cooling spray into the de-tuned duct

cleared temperatures, giving reheat a harder job. Changes to IGV scheduling and LP1 NGV area allowed the engine to operate up to limiting T3 and T6 without affecting performance elsewhere.

In reality, the engine had to cope with lower compressor temperatures than it was originally designed for and, as a result, it was eventually possible to reduce engine cost by switching to cheaper materials, particularly of front end blades, discs, etc. A more complete account is given in *Fast Jets* by Cyril Elliott.

Flight testing

In spite of the extremely tight schedules, by working feverishly round the clock the first two flight engines, which had been built and tested at Derby, were delivered to St Louis in May 1966 – more or less on time.

The ground runs at St Louis were notable firstly because they were run in a heatwave – 13 consecutive days at over 100°F and 95% humidity. Several problems were encountered and overcome – on the first engine run in the aircraft, foreign object damage was caused to the compressors of both engines by aircraft manufacturing debris (rivet mandrels, etc) due to the provision of inadequate green run screens by McDonnell. In addition, excessive oil emission from the breather earned the engine the name 'greasy pig' or 'puff the magic dragon', and made standing on the wing to carry out maintenance an extremely hazardous operation. It was suggested by the Rolls-Royce engineers at St Louis that this was caused by the angle at which the engine was installed in the aircraft (11°). This was initially ridiculed by the design office in Derby but proved when the engine was tilted to 11° in the small McDonnell mobile test stand by the simple expedient of raising the front of the stand on railroad ties (sleepers). A design modification to one of the bearing seals eventually fixed the problem but not before the problem was demonstrated at Edwards Air Force Base to Cyril Elliott and John Clarke where it ruined the latter's clean white shirt!

An interesting problem occurred on initial ground runs when the pilot slammed the throttle to idle and the engine shut down altogether. This was traced to the cambox on the engine in which the steel cam was geared up by a ratio of 3:1 to enable all the engine and reheat functions to be accommodated within a pilot's throttle quadrant of only 70°. When the throttle was slammed shut and hit the stop, the rotational energy built up in the cam was sufficient to stretch the Teleflex cable between the pilot's throttle lever stop and the cam box and cause over-rotation at idle resulting in a 20 millisec pause in the shut-off range of the LP cock (see the cam on the fuel system diagram).
This was enough to shut the engine down. It was fixed by a stiffer cable and a lightweight sculptured titanium cam, which was fitted at first lay-up in October 1966; until then, MAC fitted a second stop.

The above problems were quickly overcome and the engines prepared for first flight, but they repeatedly surged on taxiing out and all attempts to understand why were unsuccessful. This would have been a normal development problem had it not been for the fact that 'Old Man Mac' (James McDonnell, founder of the company) had organised a massive First Flight ceremony to which he had invited the US and British media en masse (including CBS television) and many dignitaries, including the British Ambassador. The event was due to take place on Tuesday 28 June 1966 but by the afternoon of Sunday 26 all attempts to solve the problem and achieve a first flight had failed; Rolls-Royce's shortcomings were about to be revealed on nationwide television to the British and American public.

Late on the Sunday, an amplifier test set was loaded into the back of a station wagon and coupled by leads to the engine electronic control amplifier in the aircraft. It was planned that the station wagon would run alongside the aircraft as it taxied out but, before it was able to depart, the test set showed that as the ground power supply was disconnected and aircraft power took over the amplifier shut down. With no amplifier control on previous tests, gradual drifting of the VIGVs during taxiing out caused the engine to surge (a magnetic brake was supposed to hold the controller motor stationary in such an event but the magnets proved to be too weak). A quick call to Keith Hatchett in Derby (at 2.00am his time!) revealed that the amplifier was protected by being shut down in the event of a power spike – all that was needed to restore control was to open and shut the circuit breaker for each amplifier after switching over to aircraft power; these circuit breakers were near the pilot s left shoulder so they were identified with a piece of sticky tape. With great jubilation all round, the aircraft successfully flew on the Monday and the ceremony on the Tuesday was a great success. It was a very close run affair. Later on the amplifier protection circuit was modified, and the VIGV controller motor brake magnet strengthened to prevent VIGV drift in the case of a power or amplifier failure.

Figure 55
Sir Patrick Dean, British Ambassador, and Adrian Lombard, Director of Engineering R-R, in front of a Rolls-Royce Phantom II (kindly loaned by Bill Small) in front of a McDonnell Phantom F4

Once the above problem had been fixed, the in-flight operation of the engine behind a supersonic inlet was relatively trouble-free, thanks to the precautions taken during design and the ATF testing of the engine behind a series of inlet spoilers – a situation which came to the attention of the Americans involved in the flight testing of the TF30 in the F111, where surging was the order of the day in all areas of the flight envelope, including approach, with little signs of improvement whatever was tried. A NACA deputation from Wright Field arrived in St Louis to try and understand why the two aircraft/engine types should behave so differently. They were told that, based on their experience of the operation of compressors behind the distorted inlets of engines buried in the wing and of VTOL aircraft, Rolls-Royce had developed a measure of distortion called DC60 which was used to evaluate the behaviour of compressors and a method of testing behind spoilers to overcome the problem. They left and nothing more was heard, but it is quite likely that what they learned was influential in the subsequent decision to ask Rolls-Royce and Allison to propose a Spey variant for the replacement of the PW TF30 in the LTV A7 (see chapter on the TF41). The PW engine in the A7 was experiencing surges in all regions, but particularly on catapult takeoff.

A team of Rolls-Royce, Lucas and Dowty engineering and service personnel was stationed at Edwards Air Force Base where the major flight tests were carried out. These included aircraft/engine handling trials around the envelope and service trials such as refuelling, etc. The major problem, however, concerned high-speed performance – as described below.

After a few flights the pilots complained that the T6 shaping (by T1), which was intended to ensure a constant maximum HP turbine blade temperature, meant that indicated maximum T6 varied round the flight envelope and they had no way of knowing whether it was being exceeded or not. Dave Whittock prepared a chart for the pilots to consult when flying showing how T6 should vary with altitude and Mach number but Bud Murray, the Project test pilot and others (including Wilf Ewbank and Peter Kerry) convinced him it was not practical so T6 shaping was scrapped; this may have been a contributory factor in the short blade HP turbine blade lives in early service aircraft.

Prior to carrier trials on the USS *Saratoga*, the wave off and bolter performance of the F4K was assessed at Patuxent River, the US Navy Flight Test Center. For a two-engine deck approach with full flaps, the engines were throttled back in 12th stage BLC. At touchdown in preparation for a bolter if the arrester wires were missed, or in the event of a wave off, the engines were opened to Military (reheat unlit) rating, and BLC switched to 7th. A single-engine approach was obviously at a higher thrust level than for two. For wave off and bolter the engine was opened up at maximum (reheat lit) rating and BLC switched to 7th. For carrier landing rapid reheat light up and acceleration were essential. When the engine was not in reheat, the fuel manifolds were empty and had to be filled by the reheat pump on light up. Following the trials a modification was designed to achieve maximum reheat in two seconds from selection by priming the manifolds with LP fuel, but it was not bought by the MoD.

Another modification which came to nothing was the Moog valve. A potentially disastrous situation could arise if reheat failed leaving the final nozzle on pre-open on catapult takeoff; this would result in a low thrust for a critical takeoff. To deal with this, a valve sensing a sudden drop in jet pipe pressure was arranged to shut off the reheat system, and thus close the nozzle down to Military (reheat unlit) power. The valve was made by Moog and was set up on an engine on the test bed at St Louis. Jim Hansen, the very droll MAC Project Flight Test Manager, described what happened (expletives deleted):-

"We ran the engine up to full reheat and the valve (with no triggering drop in jet pipe pressure) shut the reheat down.

We fitted a second valve to the engine and ran up to full reheat but it still shut the reheat down!

We disconnected the wires to the valve but it still shut the reheat down!

We took the valve off and threw it over the fence but it still shut the reheat down!!!"

There is no need to state the fate of the Moog valve!

Overall performance of the F4K and M

The performance of the Spey in the Phantom was initially very disappointing with fuel consumption being too high and aircraft climb and acceleration being well below specification. Modifications were introduced to improve both the basic engine SFC and the reheat system thrust. These culminated in the Red Standard engine, which amongst various improvements provided specification maximum reheat thrust at mach 2.0 but still fell short at mach 1 due to buzz problems. However, the predicted maximum speed of the aircraft was not achieved and this led to the inevitable thrust versus drag debate. A more comprehensive report on the various performance problems is given in Cyril Elliott's excellent book *Fast Jets*, No 5 in the Rolls-Royce Heritage Trust Technical Series.

Service operation

The first aircraft to be delivered to the UK was XT 597, one of the aircraft used in the carrier trials and it arrived at A & AEE's Naval Test Squadron in September 1967. The first seven production aircraft were delivered between June and October of 1968 and one of these made its first public appearance at Farnborough in September. Wilf Ewbank, who was Rolls-Royce Service Representative with the RAF Phantoms from 1968 to 1977, has the following comments:-

"In my time with the engine, the new-to-Spey equipment performed very well with the exception of surge on the selection of reheat – this was cured by deletion of the sensing valve and BLC on needle in about 1975. This was tried at Edwards in 1968 but vetoed by the Performance Department because of the thrust loss at maximum reheat on 12th stage bleed; understandable for carrier operations but okay after 1974 when the RAF took over the Navy's F4Ks. The jet pipe, gutters and nozzle, etc, all did very well mechanically.

The biggest problem was the HP turbine blade starting with the six-hole forged, followed by the seven-hole film-cooled, but not until we got the DS blades did we get a reasonable life. The six-hole blade lasted about 25-30 hours in service.

Engine health monitoring became a way of life on the Phantom and introduced some firsts:-

- *The oil system was equipped with a main chip detector and one for each main bearing chamber plus another for the HP gearbox. By checking every 10 flights or so a tendency could be built up with more frequent checks if a failure seemed imminent. Debris was examined by microscope (x60) and seal, roller or ball bearing debris could be identified giving an indication of where the pending failure was.*

- *HP turbine seal debris plus a 1% drop in NH at takeoff (reheat unlit) was a sure indicator of a failed HP turbine blade.*

- *The Phantom Spey was the first engine to have module changes which helped with the HPTB problem and provided a lead in to the Adour.*

- *Two F4Ms performed the longest ever non-stop flight by a RAF fighter by flying from Coningsby to Singapore in 14 hours, 35 minutes in 1974."*

In the event, the government did not make the money available to modify HMS *Eagle* for Phantom use and so the F4K was never required to operate from the 150' waist catapult. This had been one of the main reasons for selecting the Spey with its higher thrust capability rather than the incumbent J79. The F4K served with the Royal Navy until the aircraft carrier *Ark Royal* was eventually taken out of service in 1978, just four years before the Falklands conflict.

Following the departure of the *Ark Royal*, the F4Ks were transferred to the RAF who had received their first F4Ms in July 1968. The RAF operated these aircraft until they were finally withdrawn from service in October 1992.

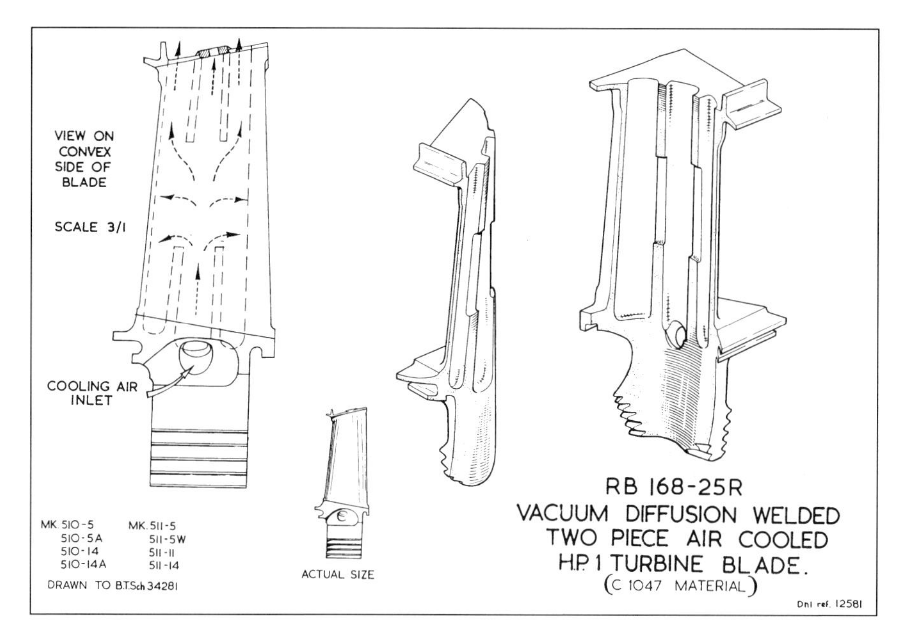

Figure 56

Figure 57
Phantom F4's of the Royal Navy (plus one from the US Navy) and Buccaneers on the Deck of HMS Ark Royal

Chinese engine

In 1972 the People's Republic of China expressed interest in a licence for the Spey. With no little surprise, it was learnt that they meant the military Mk 202, and not the civil Mk 512, which they had been operating in their Tridents. The agreement was eventually signed in 1975 for the Mk 202C. Fifty engines were manufactured in the UK. The Xian factory, which long ago had produced Nenes, was converted for Mk 202 production. In 1979 the first Chinese-built engine was brought to Derby and satisfactorily completed a 150-hour type test. Meanwhile, a team of Chinese engineers was stationed in Derby for a comprehensive training programme in technical engineering as well as manufacturing, service, etc. A formidable quantity of manuals and training documents were shipped to China.

In 1988 a prototype Chinese aircraft was said to be powered by two Mk 202s, but this was later denied. More recent information confirms that the Mk 202 was installed in the XAC JH-7 and that it has flown.

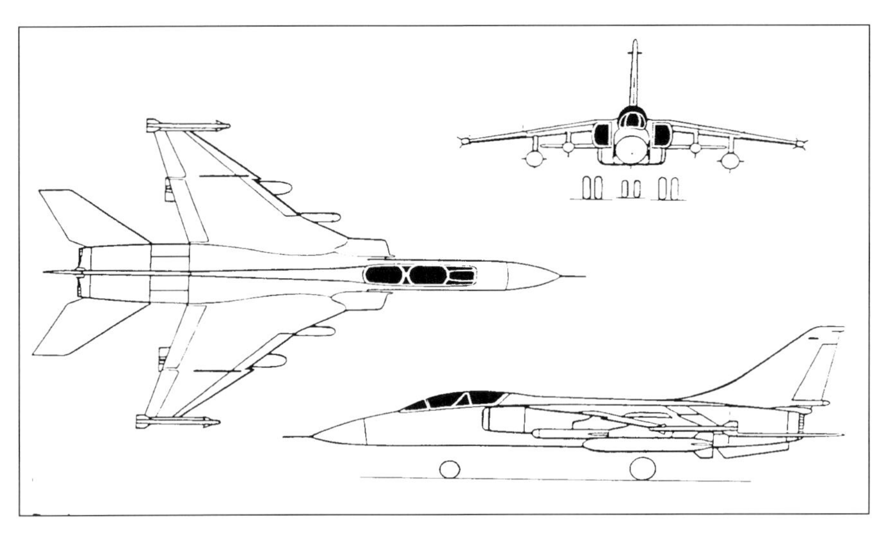

Figure 58
The Chinese XAC JH-7 aircraft powered by two RB168-25R's

Thrust SSC

Probably the most interesting and challenging installation of the reheated Spey was in *Thrust SSC*, Richard Noble's land speed record car, which not only established a new record at Black Rock Desert in the USA on 15 October 1997, but in the process twice exceeded the speed of sound. The car, driven (piloted?) by Andy Green, an ex-RAF pilot, used two ex-Phantom Mk 202 reheated Spey engines.

Figure 59
Richard Noble's 'Thrust SST' car which was powered by two RB168-25R engines to two world land speed records

Figure 60
A cut-away view of the Allison / Rolls-Royce TF41 engine showing the 3 stage LP - 2 stage IP compressor arrangement. The remainder of the engine, apart from installation features, was RB 168-25 based

Chapter seven: TF41

The Rolls-Royce/Allison engineering and marketing collaboration began in 1958. After initial setbacks of failing to be selected for the TFX (F111) and Boeing 727, Spey fortunes were at last to change. The Vietnam War was raging in the mid 1960s, and the US Military were in need of another engine source.

The Ling-Temco-Vought Corsair II A-7A was a low-level subsonic strike aircraft, land based for the USAF and carrier based for the USN. It was fitted with the 11,350 lb thrust PW TF30-P6, a reheat version of which was in the F111, and giving many problems, particularly surge either when being launched by steam catapult or due to inlet distortion. The USAF was planning an updated version, the LTV A-7D, and requested a proposal based on the Spey. The requirements were:-

1. Maximum thrust increase over the TF30 to increase payload and manoeuvrability.
2. High tolerance to inlet distortion, especially during weapon-firing.
3. Inlet airflow not to exceed the TF30, a 25% increase over the Spey –25R.
4. Existing Spey –25R parts to be used, where possible.

The RB168-62/912-B3 was submitted and, in 1966, a contract was awarded for what became known as the TF41-A1. The A-7D was a tactical fighter, retaining the folding wings of its naval predecessors for storage at combat airfields. In July 1968 the USN placed an order for the slightly uprated TF41-A2 in the light attack bomber LTV A-7E, which would be carrier-based. This was further followed by other land- and sea-based models, some of them two-seaters. The Corsair was designed to be vicious rather than pretty, and was known as the Short Little Ugly Fellow. The TF41 versions were christened Super SLUF.

Design and development of the TF41 was a joint responsibility. Production engine manufacture was split roughly 50/50, with Allison providing assembly, test and service support.

Figure 61
The LTV A7 powered by a single Allison/Rolls-Royce TF41 on carrier approach

TF41 design

The obvious way to increase the thrust of the –25R was to use the permitted +25% of airflow. The larger LP compressor would also need an increased pressure ratio in order to pass a significant part of the extra flow through the existing core. A bypass ratio of 0.76 was selected, almost half way back to the original Mk 505.

In the years since the RB141 was designed, research work had been carried out at Derby and Indianapolis on transonic compressors without inlet guide vanes and considerably higher flow/inlet area, achieving satisfactory efficiencies. This work led to the TF41 LP compressor, a free vortex design with axial air exit angles from the stators, lower blade aspect ratios, lenticular blade profiles with maximum thickness at mid chord, and much higher tip velocities giving more pressure ratio per stage. In 5 stages, it blew 3.4 compared with 2.76 on the –25R. In order to prevent the cold/hot mixing pressure ratio getting even higher, the bypass air

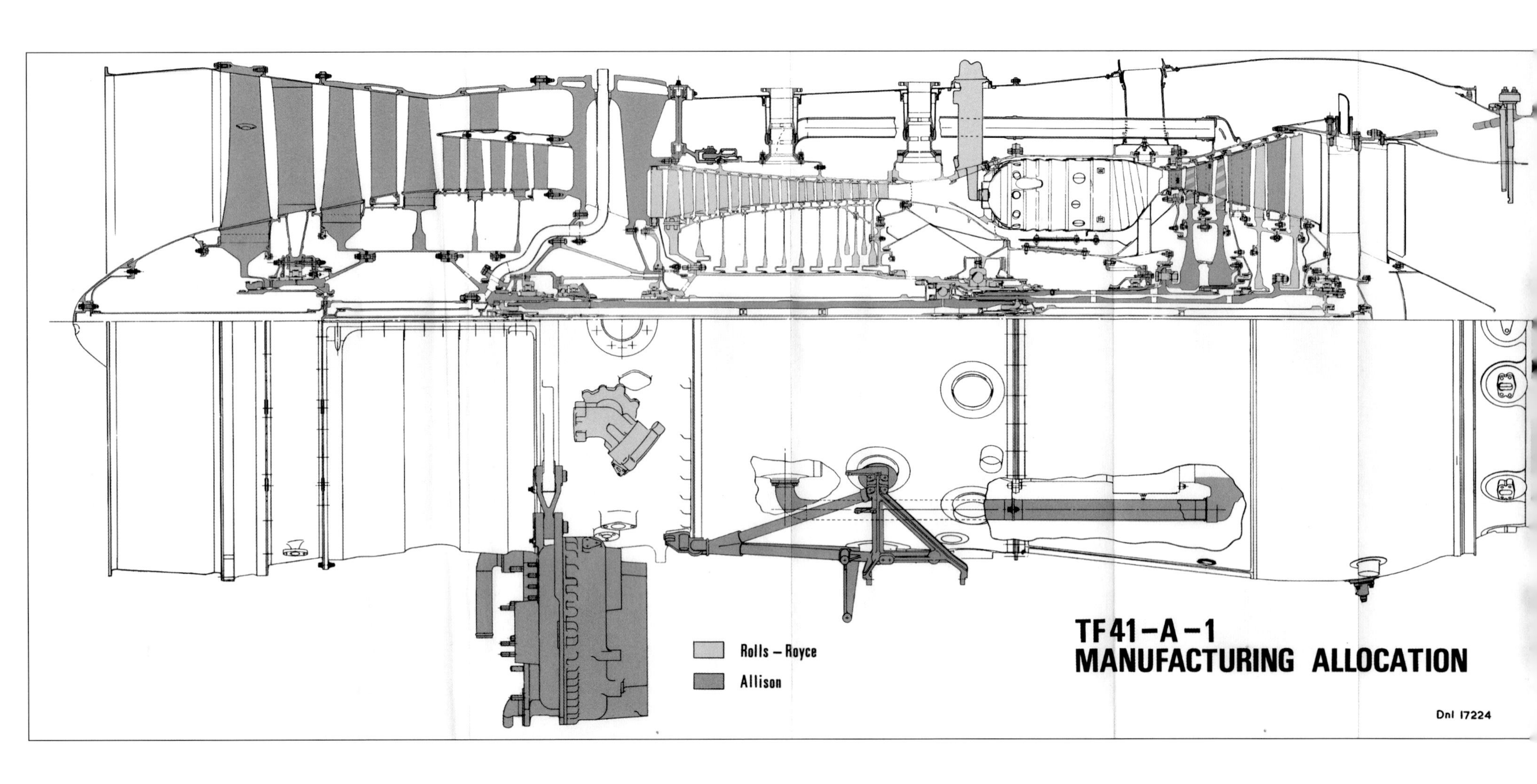

Figure 62
TF41-A-1 Manufacturing Allocation

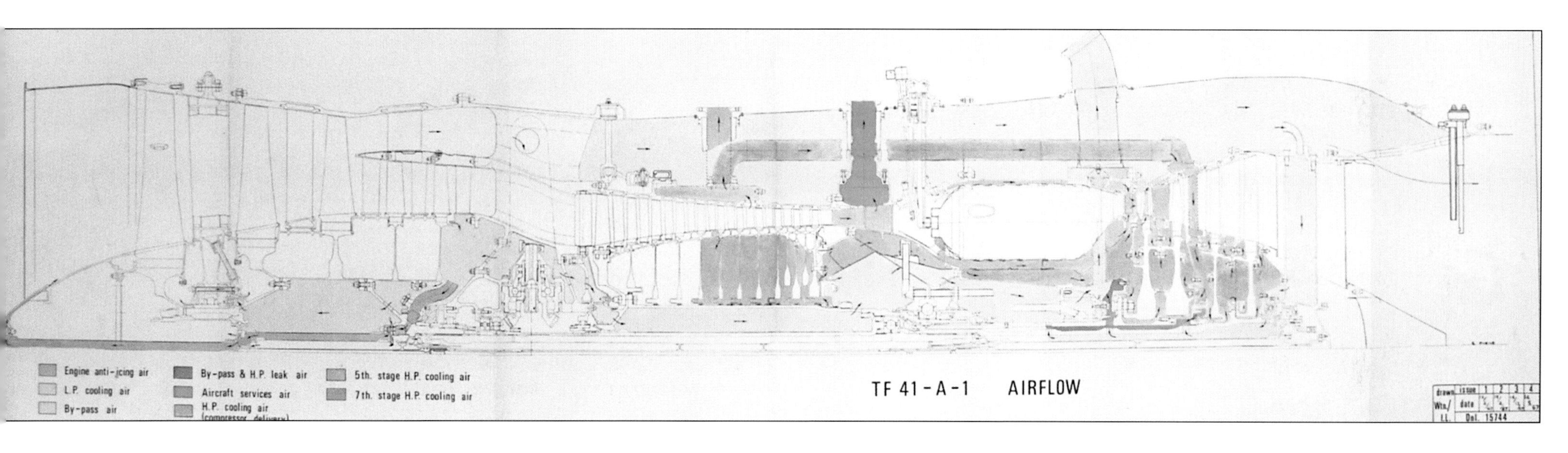

Figure 63
TF41-A-1

was taken after stage 3, and the rear 2 stages cropped to form an LP/IP design. It was sized at 37.5" diameter (other Speys were 32.45") to give an airflow of 258 lb/sec, but with an LP1 rotor tip speed of 1450 ft/sec compared with 1150 on the earlier Speys, so that the rpm was actually about 7% higher. This mitigated somewhat the harmful effect of the larger compressor on LP turbine efficiency and LP shaft torque. Even so, the torque was 75% more than the original Spey, and double the increase of the –25R. To compensate for this, the life requirement was less than for civil engines, and the proportion of time spent at full throttle was less than the –25R.

Changes from the –25R to the TF41-A1 were:-

- The LP compressor rotor was overhung ahead of the front squeeze film roller bearing, which was supported by the LP1 stator blade.

- LP cooling air was taken through the struts at the front of the bypass duct. In the absence of LP IGVs, the nose bullet was the only engine part needing anti-icing. This air was taken from the HP7 manifold, across the BPD, back through the struts and then up inside of the LP shaft.

- The swan neck duct between IP and HP compressors was the most severe at the time, but nothing compared with more recent engines.

- With a higher pressure at HP inlet, the 12th stage had to be knocked off to prevent the overall pressure ratio exceeding the –25R value of 20.0. This led to a longer exit diffuser of shallower angle.

- To achieve this matching, the HP and LP turbine capacities had to be increased by 12% and 6% by skewing their front rows.

- The maximum thrust of 14,250 lb was flat-rated to ISA+10°C (77°F), corresponding to a TET of 1425°K as on the –25R. The forged HP1 turbine blade was a seven-hole type H in N108, and HP2 in N118.

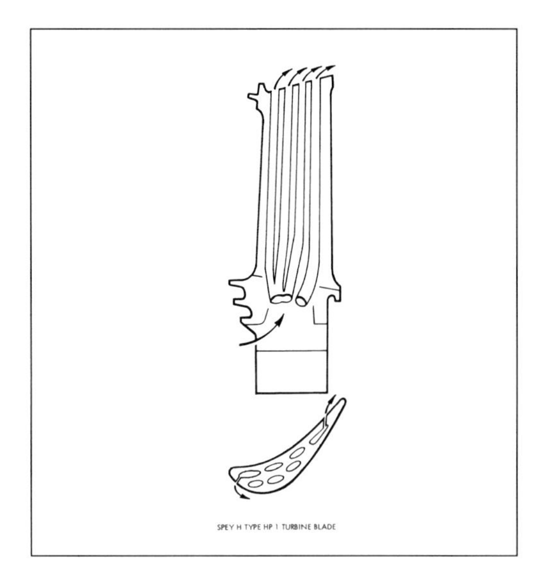

Figure 64
The revised design 7 hole HP turbine blade

- With an 11-stage HP compressor, the HP turbine work was reduced by 9% or 37°C, and in order to avoid an increase in the uncooled HP2 blade temperature the work split was adjusted by loading stage 1 and increasing the reaction of stage 2. This led to an increased number of HP2 NGVs.

- LP turbine discs were cooled by HP7 air instead of HP delivery in order to reduce the SFC penalty and improve HP compressor handling. The air was ducted from the HP7 manifold and then through the LP1 NGVs, which were cored, but not themselves cooled.

- The TF41 had six exhaust unit struts spaced as for 10, fitted with aerodynamic fairings to deal with the 14° exit whirl from the more heavily loaded LP turbine.

- The Corsair jet pipe was two diameters long, giving a satisfactory mixing gain with an annular mixer.

The fuel control system was similar to the Lucas CASC on the –25R with reheat and BLC deleted, and features added to conform with US MIL Spec for the Corsair:

1. The flat rating to ISA+10°C was achieved by an NL/T1 control, which limited the engine airflow. The P3 limiter was set higher.
2. For a single-engine aircraft, there had to be an emergency manual fuel system to take over the duties of the main fuel control CASC in the event of a CASC failure; in addition, there was a dual HP pump.
3. A gun and rocket fuel dipper was provided to avoid handling problems arising from inlet distortion and the effects of entry of unburnt propellants. The fuel was temporarily dipped prior to firing, but not enough for flame-out.

For the US Navy, maximum rated thrust of the –A2 was increased from 14,250 to 15,000 lb by raising the airflow limiter slightly; this flat-rated the engine at ISA (15°C) instead of ISA+10 (25°C). Although this did not increase the maximum TET, it did result in a more severe usage. This was compensated for by a reduction of the scatter between engines by various trimming procedures. Other small changes were introduced, recognising the role of the aircraft; better corrosion protection was applied, and some compressor discs were thickened. The development programme was quickly completed in late 1968/early 1969 and production deliveries commenced later in 1969.

It was clear from our presentation to the US Military prior to signing the contract, that they were very concerned about the engine s ability to stand the severe inlet distortion, particularly during steam catapult launch. We assumed this stemmed from reported problems on the TF30, and there was some anxiety about its LP/IP compressor design. By then, the Conway Co42 was in service with an LP/IP and we could claim that we understood its handling behaviour (see chapter on Phantom Spey). We pointed to LP/IP rig testing with varying bypass ratio and spoilers designed to reproduce the same distortion parameters as in the aircraft. These showed a degradation of IP surge margin when tested in tandem with the LP, due to interaction. To be on the safe side, the TF41 was designed with higher surge margins than previous military Speys on the LP, and an even higher (40%) margin on the IP.

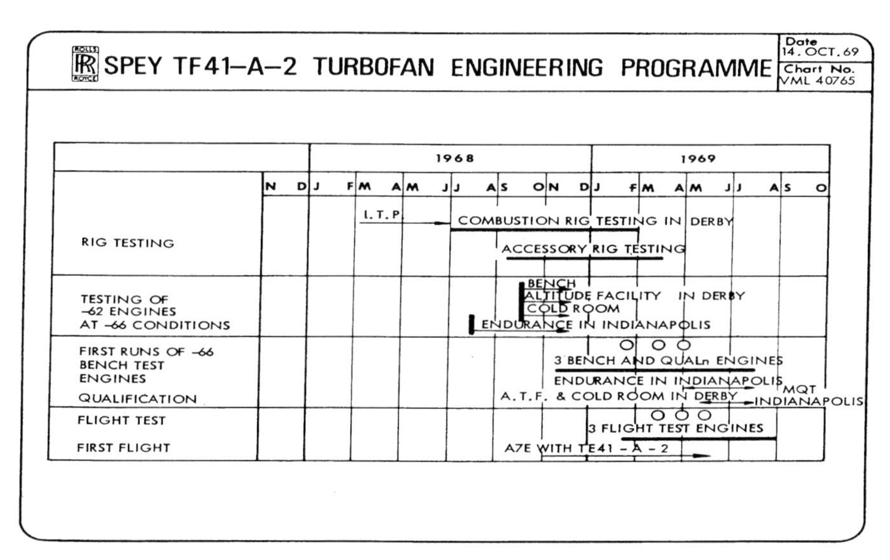

Figure 65
The TF41 development programme – again a very short/programme

ROLLS ROYCE

Comparison of Rolls-Royce Engine Compressor Surge Margins

	SPEY					CONWAY
	TF 41		RB.168-1	PHANTOM RB.168-25		
	TARGET	ACHIEVED	BUCCANEER	UNSPOILED	SPOILED	
LP	30%	28%	10%	22.5%	DC_{60}-0.21 15% -0.4 11%	10%(25%)
IP	40%	41%	—	—	—	25%(10%)
HP	32%	36%	25%	28%	DC_{60}-0.81 19%	35%

SURGE MARGIN EXPRESSED AS $\left[\frac{(R_{Surge} - 1)}{(R_{Operating} - 1)} - 1\right]$ 100%

MARGINS IN BRACKETS ARE FOR LP - IP IN TANDEM

69 J SML 13325/1

Figure 66
Surge margins on various R-R engines showing the high margin designed for and achieved on the TF41 IP

Development and service

Well before 1966, there was an exchange of design and development engineers between Derby and Indianapolis. Following the TF41 contract, larger teams were established to carry out the very short programme. The first development engine ran two weeks ahead of schedule at Derby in October 1967, followed by four more in the next five months, by which time the first three qualification engines ran in Indianapolis.

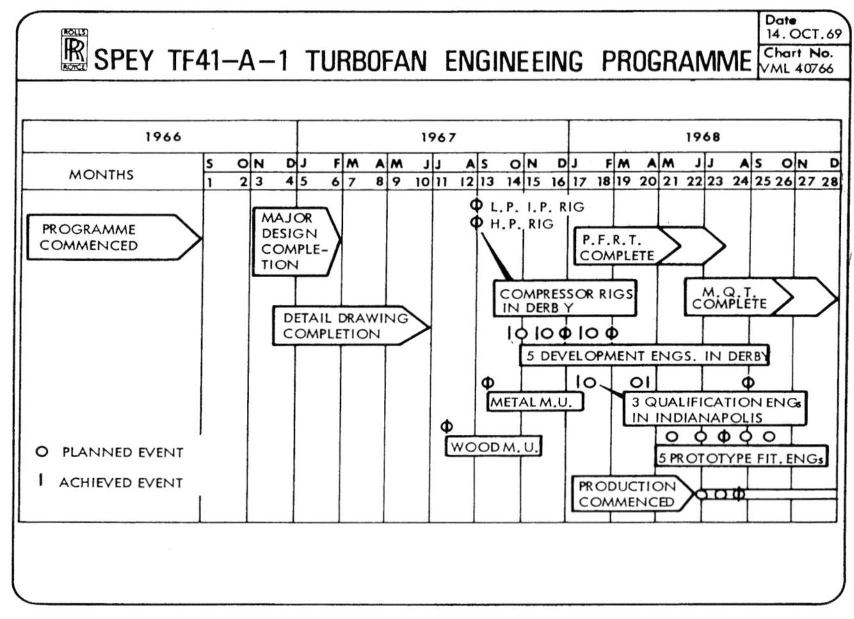

Figure 67

In common with other first runs, the engine was not right first time. This time it was excessive TET at a thrust, which stubbornly resisted attempts at improvement. The first three engines ran about 60°C hotter at a thrust than brochure, a main cause of which was a core flow up to 8% too low. The standard treatment for this was to increase the LP turbine capacity, and to begin with an expected improvement was achieved by crimping the LP1 NGVs. However, further increases by skewing appeared to go the other way, which was attributed to the special TF41 design changes. We were reminded of the struggle on the Spey –1. The temptation to force more air through the core by sacrificing IP compressor surge margin was resisted, and further attention paid to opening up the LP1 NGV and blade, but even so the design flow was never fully achieved.

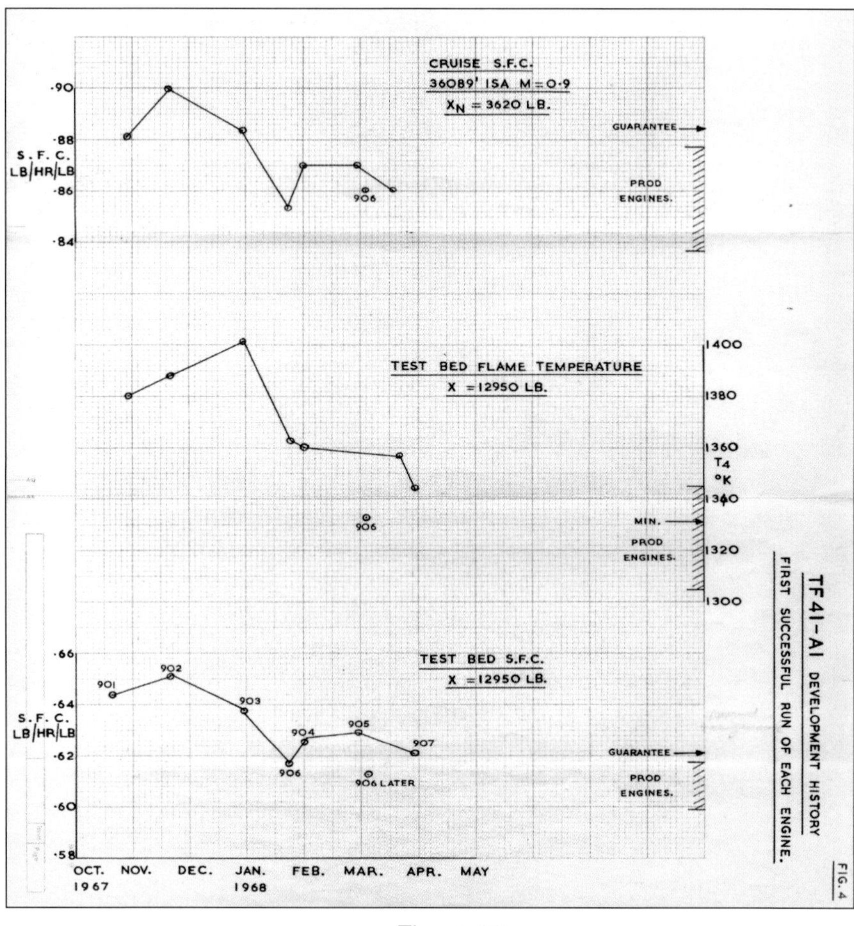

Figure 68
Chart showing the progress of improvements to the TET and SFC during the development programme

Over a period of seven months, thanks to the incorporation of 'brasso' modifications, the TET excess was eliminated and, at the same time, the target flight SFC was met. A better-than-expected HP compressor surge margin enabled the 'Fokker' bleed and IGV scheduling to be used, which greatly improved the low speed SFC at loiter.

Model Qualification Tests commenced in June 1968 at Indianapolis in their air-conditioned test stand, which could provide air at 54" Hg (1.8 atmospheres) and 207°F (97°C). The engine performed satisfactorily except for one or two problems, which were fixed by redesigning the LP1 compressor blade clapper and silver-plating the blade roots. These required further clearance on a penalty run.

A large proportion of early development testing was carried out behind a fibreglass model of the aircraft intake duct and specially developed distortion simulators. These showed that handling was satisfactory. Flight-testing at Edwards Air Force Base and LTV Dallas followed in autumn 1968. These concentrated on engine performance, handling at +3g to -1½g and M61 gun- and missile-firing. The only sign of malfunction was a slight pop surge when missiles were fired simultaneously. Since the engine recovered immediately without any pilot action, the test was deemed to be satisfactory. The ultimate handling demonstration was at Patuxent River, consisting of gross weight launches with fully degraded catapult. The engine was surge-free with maximum steam ingestion, in contrast to violent surges and flame-out on the TF30.

The A7D entered service in 1969, and was in combat in SE Asia in 1972. The earliest A7Es were fitted with TF30s, followed by the TF41-A2, which entered service in Vietnam, performing exceedingly well. In contrast, turbine failures were being experienced in non-combat training missions after only 200 and 400 hours. The guilty component was the inner spacer between the HP1 and 2 turbine discs, which was suspected of cracking due to gyroscopic loads suffered during aircraft manoeuvres. These loads could only be reproduced on a rotating test stand, and spacer stresses measured by strain gauges. In an incredible 30 days, Allison designed, built and ran a test stand capable of rotating about a vertical axis. The cause of the failures eventually turned out to be the operating cycle and not gyroscopic forces. On initial acceleration the spacer heated up more rapidly than the thicker discs, lifting off its location and allowing a transient vibration to build up, which was enough to cause an uncontained failure. The cure was to redesign the spacer, which Allison manufactured and fitted to all aircraft in double quick time. They certainly got things done!

By the 1980s, 1417 TF41s had been delivered and earned themselves a fine reputation, with an exceptional overhaul life for a combat engine of 1500 hours.

Figure 69

Installation of the TF41 through the rear of the aircraft. The picture also shows the long jet pipe which meant that good mixing gain was achieved with an annular mixer

Figure 70
The rotating test stand built by Allison in an incredible 30 days

Reheat demonstrators

With an eye on a wider military market, much effort was spent on the study and supporting demonstrations of reheat derivatives of the TF41. The first project was the RB168-67R at 24,800 lb, an A2 with a –25R design of reheat pipe. Rolls-Royce and Allison carried out a jointly financed exercise to demonstrate its feasibility. The 44" diameter pipe had originally been built by Allison and tested on the RB141 in 1960, and was afterwards purchased by Rolls-Royce from the USAF. The tests were run at Derby in 1969.

A maximum boost of 62% was achieved compared with a required 68%. This was not limited by buzz but by overheating of the outer vapour gutter, a condition which could have been cured by re-designed gutters, and other developments already in the –25R. Surge-free light ups were achieved with fuel flows corresponding to 40% boost without nozzle pre-open, and 32% with inlet plates giving much more severe distortion than the Corsair in flight. These tests were further proof of the excellent handling of the TF41 LP/IP compressors. Details of the pipe and the tests are given in *Fast Jets*.

In the years up to 1985, further reheat demonstrations were run by Allison in support of higher thrust derivatives. These were targeted at the F14 and other high performance aircraft, but they came to nothing.

Figure 71
An experimental TF41 fitted with a 44" reheat pipe at maximum reheat on a Derby test stand

Chapter eight: Scotland (up to the Tay) and service experience

In 1968 events in Derby took a major turn with the announcement of the launch of the RB211 in the Lockheed TriStar and the large order book that went with it. This was the breakthrough for which the Company had been striving for many years and the Spey in its varied forms had played its part in securing it. There is no doubt that knowledge of Rolls-Royce's capability had permeated throughout the USA, in particular, in the airlines American, Braniff, and Mohawk with the Spey in the BAC One-Eleven and the US services and NASA with the McDonnell Douglas Phantom Spey and the LTV TF41-powered A7 Corsair.

This sudden increase in workload meant that the decks needed to be cleared in Derby and so all responsibility for engineering, service support and assembly and most of the testing of the Spey was eventually moved to Scotland; with it went key members of the engineering and service support staff. The move of the civil Spey was delayed initially as David Huddie wanted the support of as many engineers as possible in Derby during the RB211 selling campaign. The first engine to be moved was the RB168 for the Buccaneer with design and development moving into Hamilton in 1968; the Nimrod Spey was also nominally moved at that time but, since the engine was basically a civil Spey, much of the work remained in Derby.

Giles Harvey, who had been appointed Chief Engineer for the Civil Spey in April 1968, moved in to Hamilton in September 1968 and took on responsibility for all Scottish civil engines. Working for him as Project Manager on the Spey was Colin McChesney, as Chief Performance Engineer, Alun Jones and Development Engineer Barry New. The responsibility for service support of the Spey at that time was still separate with Ken Hughes moving from Derby to East Kilbride under Bert Spufford. Eventually all design and development responsibility for the civil Spey was moved on 06 January 1969 with build and test remaining in Derby and nacelles in Hucknall. Subsequently, all engineering and service responsibilities were combined for each of the projects and everyone was moved to a new building at East Kilbride on 05 November 1971; Hamilton was closed.

The Phantom Spey was the last to go to Scotland in 1975, John Clark moving with it to keep continuity on the service support programme.

In-service problems encountered and solved

In 1968, at the time of the move up to Scotland, the civil Spey had only been in service for four years, but was already established in the Trident and BAC One-Eleven, and being engineered into the Fokker 28 and Gulfstream II. The interface with these four aircraft and their customers was very active. The basic design of the highest rated civil Spey, the Mk 512, had been completed in Derby but the final installation and contract/specification was carried out with both BAC and HS for the BAC One-Eleven and Trident 3 respectively, in Scotland. In 1969 the Spey Mk 555 entered service in the F28, and in 1970 the Nimrod entered service with the RAF. The tropical trials of the Mk 512 in the Trident IIE were carried out at Madrid in 1970 and for Trident IIIB in 1971.

Some of the major service problems and their fixes are dealt with in the following paragraphs. Though some of them originated while the engine was in Derby, most of the experience and the engineering solution to these problems was carried out in Scotland.

LP compressor drum failure:

One of the first serious incidents was the uncontained failure of an LP compressor drum on a Spey Mk 505 in a British Airways Trident 1 in Rome during takeoff in October 1970; it occurred after only 7,744 flights compared to the predicted safe life of 14,500 flights, ie, half the predicted life. The subsequent investigation, involving recordings of NL on a series of flight operations in BEA Tridents, identified the cause as the effect of the aircraft auto-throttle on engine cycles per flight, which were significantly in excess of those assumed in the life calculations for the drum. In effect, the LP drum was experiencing about 2.0 stress cycles per flight compared to the figure of 1.035 used in life prediction. As a result, the predicted safe cyclic lives were reduced for Trident operation.

HP compressor stage 10 disc failures:

The introduction of the Spey 510/511 ratings for American Airlines' BAC One-Elevens led to a number of service problems; prime among these was the failure of stage 10 discs. About a dozen failures occurred mainly in American, Mohawk and Braniff and caused much concern within the airlines as it was proving difficult to solve. Eventually it was found that the problem was partly due to weakness in the original design and partly due to mal-fitting on overhaul. The original stress analysis, done by desk top calculator, had not made sufficient allowance for the bending moment caused by the centrifugal force of the two spacers hung on the rear of the disc exceeding that of the single spacer on the front of the disc. It was also shown that excessively forceful re-assembly of the stage 10 disc and spacer could aggravate the loading imbalance leading to low cycle fatigue. High life service engines were x-rayed in situ to check which ones were distorted. The solution was a more careful assembly and lower life limit, prior to the availability of a thicker disc and spacer.

HP compressor stage 2 and 3 disc neck failures:

Another serious problem, mainly with the American operators was the failure of the front HP compressor disc necks and blade dovetails. A possible cause of the problem was thought to be due to excessive rotor blade vibration due to surges on landing. A BAC One-Eleven owned by Autair was hired for two weeks and the Derby instrumentation group, led by Hedley Bullock, produced a magnificent array of equipment to install in the aircraft with half the seating removed. Blade vibrations measured magnetically during flight and landing roles indicated conditions which would cause high disc neck stresses, particularly when surge occurred during the slowing down after landing and with the thrust reversers in operation.

At the same time, Allison – who were running the TF41 programme in Indianapolis – undertook a major investigation of the HP compressor rim failures during which the Spey design history was studied. Allison's finite element programme showed that the Spey front disc necks were not in balance causing very high direct stress under cf. Direct stress measurements were made by Allison, the first ever on an HP shaft. It was clear that the high direct stresses measured added to the high blade vibrations measured on the civil Spey programme were the cause of the failures. The redesign of the disc in the civil programme plus the action to ensure use of a lower NH (from 94% to 90%) during thrust reverser operation, coupled with cancellation of the reverser and aircraft spoilers at a higher aircraft speed, contained the problem. Strengthened discs were introduced on stages 2-9 in addition to stages 10 and 11.

American Airlines parts cost guarantee and reliability demonstration:

As part of the campaign to win the American Airlines order, certain guarantees were given. One of these was the parts cost guarantee, one of the first of its kind. Adrian Lombard and Ernest Eltis had agreed $7.21/hour in January 1963 as part of the bid to secure the order. In the event, technical problems on the engine caused this to be exceeded by about 20% after allowance for escalation, etc, had been taken into account.
Another requirement imposed by American Airlines was the satisfactory completion of a 1000-hour cyclic endurance test (to their cycle) on a fully cowled engine. The test was successfully completed at Hucknall, but when the engine was stripped the HP turbine blades were holed through the cooling air passages. It turned out to be caused by running 1000 hours in front of the Hucknall paint hangar and chemical corrosion was the culprit. As Giles Harvey commented, *"Explaining all this with full colour pictures to Frank Kolk in American's Manhattan office was an experience not to be repeated"*. In the fullness of time, American Airlines were persuaded that the test on the most critical part of the engine was unrepresentative. The airline did suffer some turbine

blade failures in their service operation – on one occasion causing damage to some cars in a car park – but these were associated with a lock-plate problem.

HP compressor variables seizure:

The BAC One-Eleven's in service with Aloha Airlines in the Hawaiian Islands had a 20-minute stage length and to make matters worse every takeoff was in a mixture of salt spray and coral dust. Added to that, the flight route was frequently over a quite active volcano giving high sulphur deposits. It was not long before the HP compressor variables seized up and material changes had to be introduced; the turbine blades also suffered severe sulphidation. This was a very arduous operation as Boeing also found out when one of their Aloha-operated 737s suffered the loss of a section of fuselage skin at very high cycles.

Funny Sounding Noise – FSN:

A particularly difficult problem, which caused a great deal of agony, was the FSN, or Funny Sounding Noise, on the Spey Mk 511 in the Grumman Gulfstream II aircraft. It was only present in certain aircraft and it was not at all clear that it was solely due to the engine. At first, Grumman suggested that it was due to the fact that Rolls-Royce mounted engines on test, in a different manner from the way they were installed in the aircraft (fixed large airmeter) but, although it raised the level of vibration, it proved not to be the basic cause. The noise could not be removed by tuned dampers in the aircraft as it was non-integral, ie, it was not a function of engine rpm. Clocking of the turbines relative to the compressor sometimes made the problem disappear but in other cases worsened it; the seriousness of the problem was aggravated by the fact that it became a contractual matter. It was eventually believed that it was caused by out-of-balance in the middle section of the LP shaft due to tolerances in the machining of the splines in the compressor drum causing the shaft to bow a few thou under torque load. In any event, it was eliminated by creating a clearance on the outside of the bearing of 5.5 thou making it into a type of squeeze film though it had neither anti-rotation feature nor specific oil supply. The modification could be installed with the engine on the wing instantly proving the efficacy of the FSN cure. This modification was carried through to the Tay.

Proposed re-fanned versions of the Spey:

It became apparent in the 1970s that development of the Spey market was limited by thrust capability, relatively high specific fuel consumption compared to the engines of the day and, in particular, very high noise and emission levels. The Advanced Project Design Office in Derby under Geoff Wilde looked at various re-fanned proposals for the Spey engine. The version which received the most attention was the Spey 67, which was intended for a developed version of the BAC One-Eleven, the –700. The Spey 67 (or 606) had a fan diameter of 46.9" (cf 34.25" on the basic Spey), which gave it a bypass ratio of about 2, a three-stage IP compressor and three-stage LP turbine; the core used was that of the Spey 512. This gave the engine a much higher takeoff thrust, a much better specific fuel consumption and, most significantly, a noise level which could meet FAR 36 Stage III and reduced emission levels as the higher bypass ratio diluted the hot stream. BAC marketed the –700/Spey 67 combination strongly but failed to secure a launch, mainly because British Airways did not want the aircraft.

Many further re-fanned proposals were looked at in the 1970s. One of these was the RB163-72A, fan diameter 47.4", which Grumman considered as a powerplant for the Gulfstream III, a long range, higher cruise altitude and speed development of the Gulfstream II. The engine was in competition with JT8D, the M45H-01M, the RB401-07, the CF34 and the Spey Mk 511-8 (as used in the Gulfstream II). Eventually, the Spey 511 was chosen as it was the right size – both the JT8D and the Spey –72A were too big and the M45H and CF34 would have needed a three- or four-engine installation – and, according to the Gulfstream charts, the Spey had the best installed altitude cruise sfc of the major contenders; it was

5% better than the JT8D. I wish we had known that in 1960 when the competition to power the 727 was in full swing!

So yet another re-fanned Spey came to nought, as did all the rest until the Tay was launched from Scotland in the early 1980s by a team under John Ashmole. This had a wide chord fan based on that of the –535E4 with a diameter of 44" (because of the wide chord fan it still had the same airflow as the proposed Spey 67 with a fan diameter of 46.9"), a three-stage IP compressor and a three-stage LP turbine as on the Spey 67C but it differed in the core, which for the Tay was based on the highly successful and reliable core of the Spey 555 in the F28 aircraft. This engine entered service in the Fokker F100 and Gulfstream IV and gave a new lease of life to the Spey. Cycle details for the Tay 611 are given in the engine chart in the Appendix.

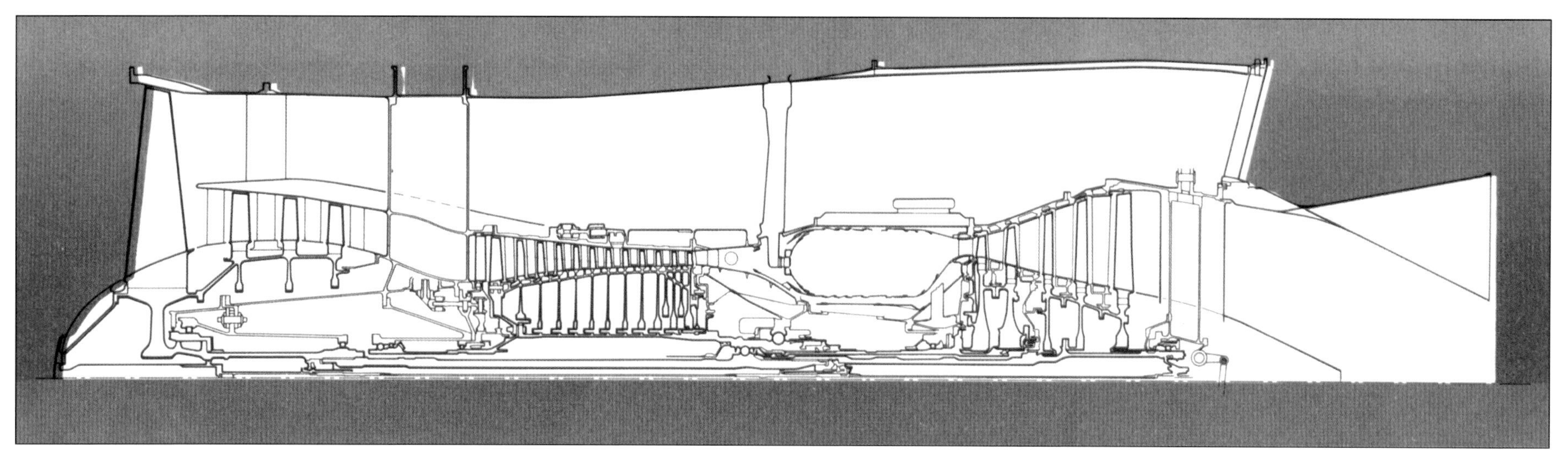

Figure 72
The proposed re-vamped Spey-67B which did not proceed as BAC failed to launch their aircraft

Figure 73
The Gulfstream III which in spite of competition from a variety of other engines, including the JT8D and Spey-67B, was powered by the Spey 511

Chapter nine: Industrial and Marine

Early aero gas turbines were adapted for industrial and marine use to exploit their flexibility. With their ability to achieve maximum power from cold very much quicker than traditional much heavier and bulkier steam plant, they were used for peak lopping in electrical power generation and power boosting in ship propulsion. Prior to the Spey, the Avon, Proteus and Olympus were used for electricity generation and gas pumping, and the Proteus, Olympus and Tyne for propulsion of naval vessels. Their roles were later expanded to continuous use, and advantage taken of their very low manpower requirement for maintenance and replacement. However, ownership was costly, and the major objective for the next generation was reduction of the specific fuel consumption, which was much higher than other methods of electricity generation or marine propulsion.

For aero engines, the all-important flight SFC depends on both thermal and propulsive efficiency. Bypass engines were introduced to reduce the jet velocity and so improve propulsive efficiency. For stationary or marine engines the bypass is an unnecessary disadvantage, as the exhaust energy in the hot stream is used to drive a free power turbine. The important parameter is thermal efficiency, which depends on overall pressure ratio (OPR), turbine entry temperature, component efficiencies, etc. Speys had a considerably higher OPR than their predecessors, higher turbine entry temperatures due to blade cooling, high component efficiencies and, therefore, a higher thermal efficiency. The most advanced Spey, the TF41, was chosen as a basis for the gas generator, minus the bypass, of course. This engine plus power turbine had a thermal efficiency of 35%, giving it a 20% better SFC than the Olympus as a marine power unit.

Fundamental changes to the aero engine were:-

- Bypass system removed by cropping the three LP compressor stages.
- Compressor casings redesigned for the above, and for new air and power offtake requirements, etc.
- Combustion chamber modified for burning a range of gaseous and liquid fuels.
- Turbine exhaust casing modified to connect with the casing of a two-stage free power turbine. The power from this turbine was used to drive a compressor for pumping, electrical generation, or a propeller.
- The flow capacity of the power turbine acts as a final nozzle for the gas generator, and the low-pressure gas is exhausted radially through large ducts.
- The engine with its inlet ducting containing flow straightening cascades, power turbine plus exhaust, shafting, gearbox, compressor, etc, was mounted on a common baseplate. Typically, a complete marine module would weigh 25.7 tonnes, of which 1.8 was the gas generator, 1.0 the inlet and 6.9 the power turbine and exhaust, with overall dimensions 24' x 7'6" x 11'.

Airflow is the most important parameter affecting power output, followed by TET. With the bypass deleted, the Spey exit airflow was 125 lb/sec, giving it a base rating of 16,900 HP or 12.6 MW. This was very roughly three times the Tyne and Proteus, the same as the Avon, and half the Olympus. Increases of TET and component performance gave the Spey an advantage in power/weight over its predecessors. This progressive advance in technology was continued with the RB211, which produced about double the Spey power, as well as further improvement in SFC.

The first industrial Spey commenced gas-pumping service in 1976, six years after TF41 service operation began in the Corsair. The first marine production version entered service in 1985.

Industrial

The basic gas generator was suitable for many different industrial applications, onshore and offshore. For gas or oil pipelines and water-pumping, the power turbine was coupled with a compressor or pump, and for electricity co-generation connected to an AC generator through a gearbox. The SK 15HE was a later plant in which the hot exhaust gas was fed into a heat exchanger, which provided steam to generate more electricity.

The base rating covered operation up to 8760 hours per year (24 hours/day) with an average of 25 starts per year, and major inspections at three- or four-year intervals. For mechanical drive, this gave an ISO power of 16,887 shaft horsepower (12.6 megawatts), and for electrical generation 12.1 MW including gearbox and generator losses. Alternatively, 12% higher maximum continuous or peak ratings could be used for 2000 hours per annum, with up to 500 starts.

Starting with a military engine protected against corrosion by virtue of its carrier operation, many features such as blade and casing materials were suitable for industrial and marine use. For new operational duties, the following design features were introduced:-

- Overboard bleed from LP stage 5 (outlet) at low speeds to counteract the effect on efficiency of a flatter working line with the deleted bypass.

- Two LP stage three blow-off valves for trip shut down.

- Variable HP IGVs and 7th stage handling bleed retained.

- Woodward electronic control system using the same parameters as the aero engine, plus exit gas pressure and power turbine rpm. Capable of operating with liquid or gaseous fuels.

In industrial and marine engines, there was much emphasis on service and maintenance to obtain high reliability and low cost of membership. Measures to achieve this were:-

- Condition monitoring on a large number of performance, mechanical and other parameters. This gave an indication of general engine health, and early warning of performance deterioration.

- Visual inspection of hot gas paths, etc, through a multitude of borescope access ports.

- Compressor washing using intake spray rings.

- Split casings for inspection and ease of servicing and replacement of some components in situ.

- The gas generator was split into five separate modules, which could be individually replaced without returning to the factory, thus minimising the time out of action.

- Facilities for quick engine removal and replacement. A change-out took 4½ hours.

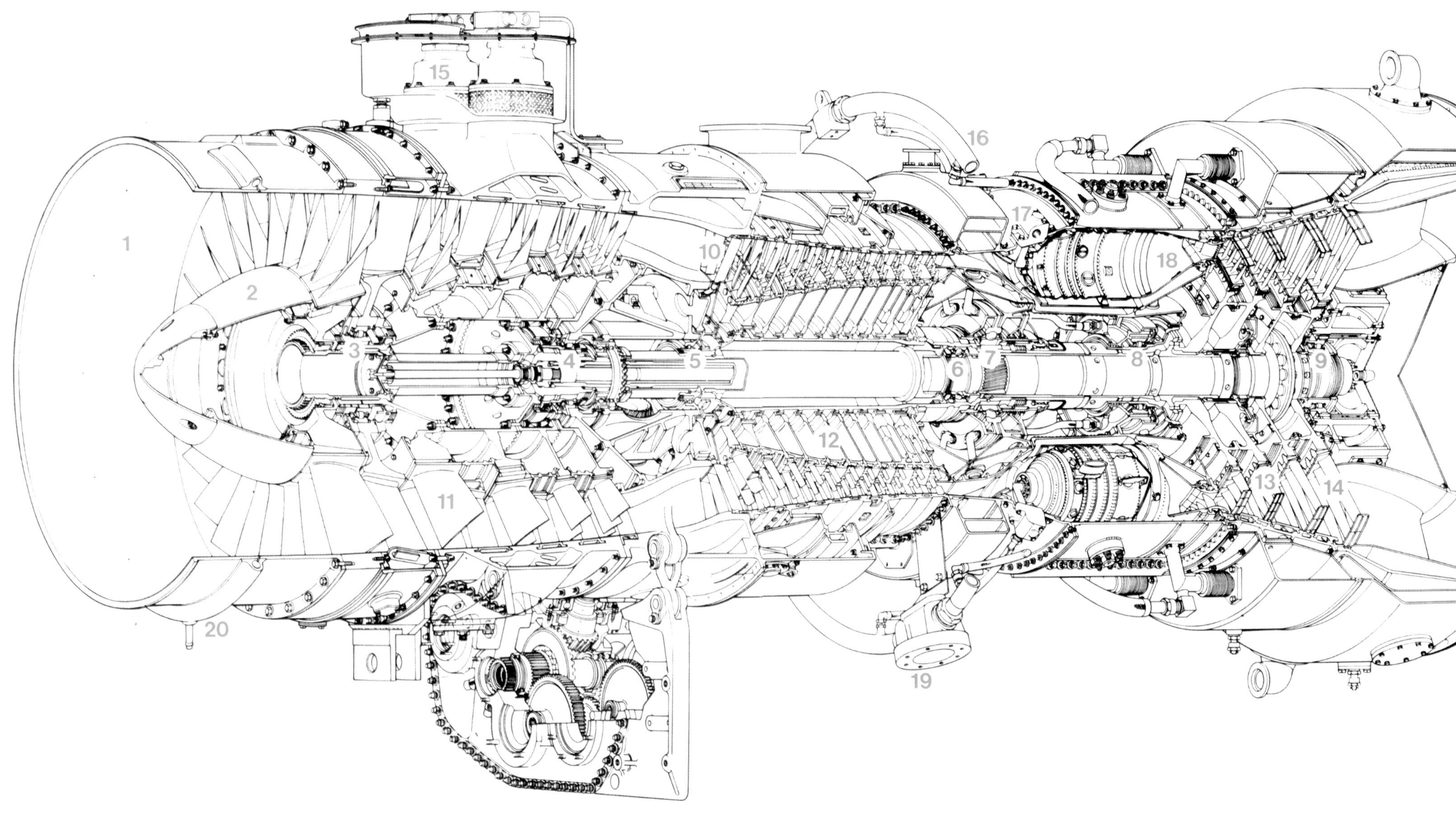

1 Air intake extension
2 Nose fairing
3 LP compressor front bearing
4 LP compressor rear bearing
5 HP compressor front bearing
6 LP thrust bearing
7 HP thrust bearing
8 HP turbine bearing
9 LP turbine bearing
10 Variable inlet guide vane
11 LP compressor
12 HP compressor
13 HP turbine
14 LP turbine
15 Bleed valves
16 Gas fuel manifold
17 Gas fuel burner
18 Combustion chamber
19 Gas fuel manifold inlet
20 Compressor crank/soak wash spray ring

Figure 74
The Industrial Spey

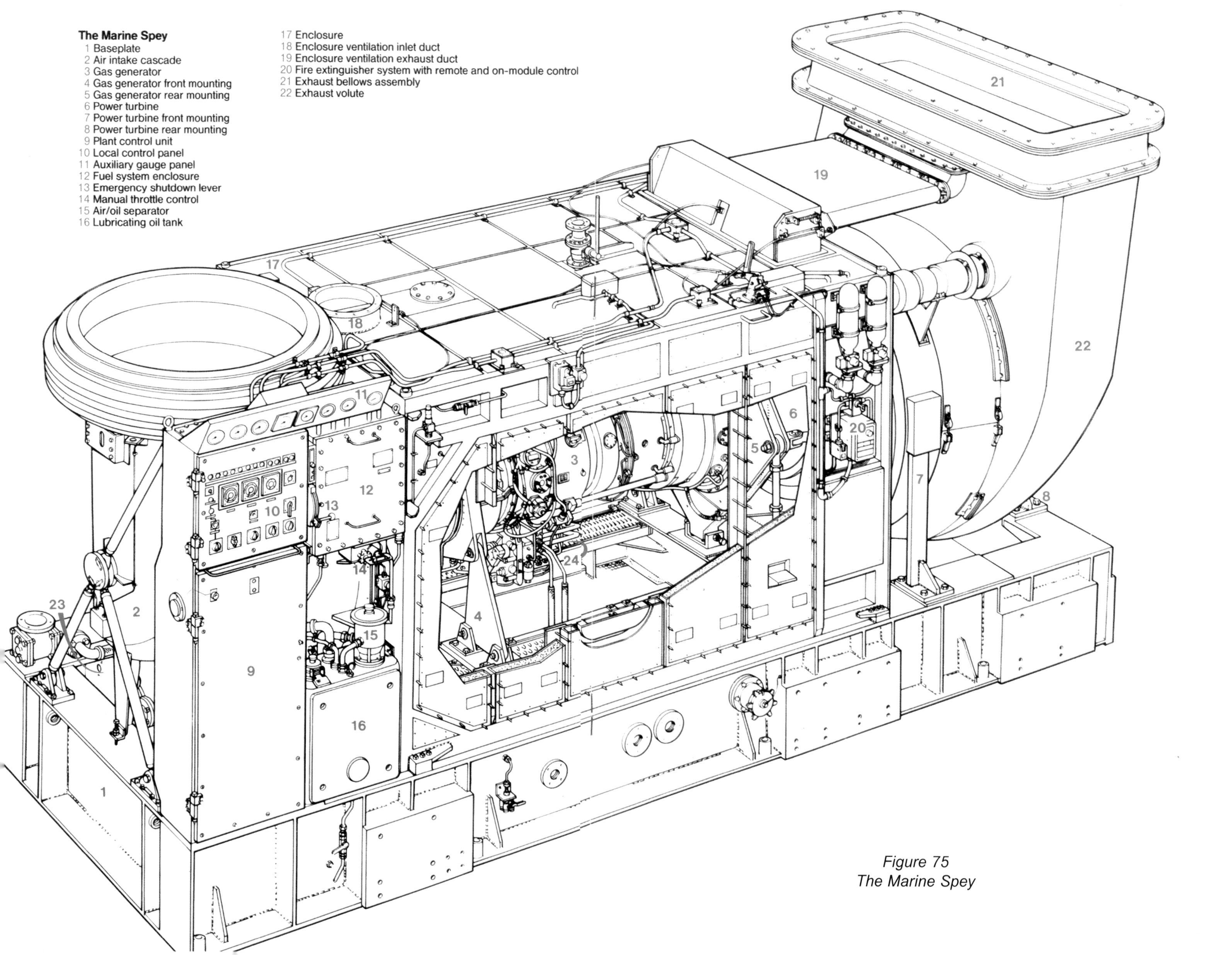

Figure 75
The Marine Spey

Marine

Already in industrial service, the marine version was chosen to meet the Royal Navy power requirement of 12.75 MW as well as its 20% SFC advantage over earlier aero engine derivatives. The first marine Spey, the SM1A was conservatively rated at 12.75 MW maximum continuous with a capability for a 14 MW sprint. To obtain early operating experience, it replaced the two larger Olympuses in HMS *Brave*, a Type 22-07 frigate. This was a COGOG (**Co**mbined **G**as turbine **o**r **G**as turbine) package (Spey or Tyne), but later Type 22s were altered to COGAG (Spey and Tyne) which increased ship speed. By 1989, 112 engines had been sold or ordered for five classes of frigates and destroyers for the UK, Japanese and Netherlands navies, with four Speys together or paired with Tynes, Olympuses or diesels. In addition, five were used in China for power generation.

Resulting from industrial experience, some changes were made to blade angles, numbers and annulus shape of the cropped LP/IP compressor. The benefits were increased part load efficiency, improved surge line and better distribution of stage pressure ratios. Forged HP turbine blades were replaced by vacuum cast to ensure adequate corrosion resistance.

Operation requirements for warships include no visible exhaust and low infra-red emissions over the range of power levels. Anti-smoke modifications were not enough, and idling combustion efficiencies unacceptable. The duplex burner was replaced with a simplex injector and reflex airspray burner, which had two recirculation zones; one for low power operating at near-stoichiometric for high efficiency, and the second at high power for low smoke emission.

The power turbine was designed for multi-engined installations of fixed gear ratio and propeller pitch. The engine would be required to produce a given power over a range of shaft speeds depending on the number of engines in operation. The engine change unit was the gas generator only, the power turbine being designed for the life of the ship, and hopefully permanent. It was, therefore, a conservative robust design. Clockwise or anti-clockwise rotation was available on twin shaft ships, permitting symmetrical reduction gear design. The two turbine stages were overhung with all the bearings to the rear, with their supports through the cooler exhaust gas. The two journals, central thrust and an auxiliary reverse thrust bearings were all of the tilting pad type and not rolling. This avoided the need for a pressure balancing system using gas generator air to minimise the axial load.

Four gas generators and two modules were used for the development programme, one at Ansty and the other at NGTE Pyestock. This continued until 1982, followed by cyclic endurance testing lasting 3000 and 2000 hours on the complete module under simulated marine conditions. The SM1A entered service in 1985, with a declared overhaul life comparable with the Tyne and Olympus, and a target to double this value. By then, however, many navies were acquiring larger ships which needed more power.

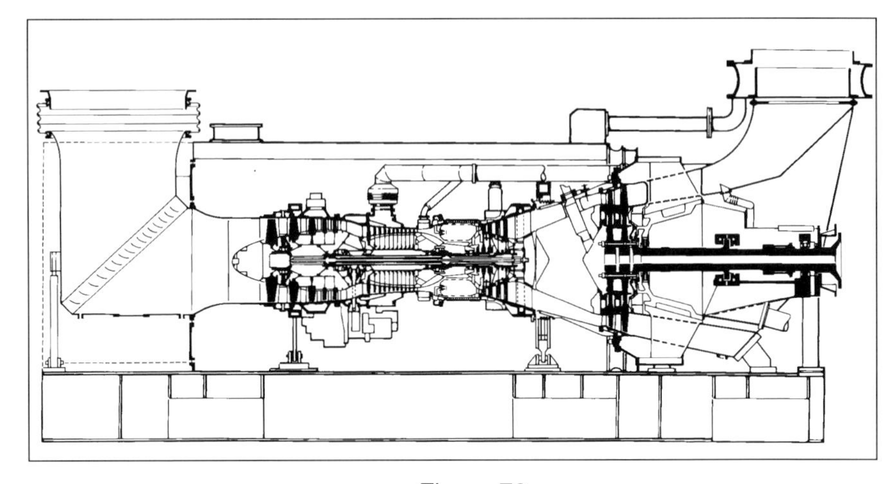

Figure 76
The uprated Spey SM1C power module was designed to fit into the same installation unit as the SM1A to ease retrofit

In 1983 a Rolls-Royce/Royal Navy study was carried out to consider the options for development to a continuous rating of 14 MW and beyond, and a lower cost of ownership. The big step in SFC had already been taken, and it was concluded that an attack on life and reliability would be more beneficial. So instead of continually squeezing more power out of the SM1A, designing for a single large power increase would give greatest scope for reducing cost of ownership, as operation would normally be at reduced power and, therefore, less arduous. The result was the SM1C designed for a 40% uprating to 18 MW continuous and 19.5 MW maximum, at 5% lower SFC.

The SM1C incorporated the later technology already demonstrated in the Tay 650 and RB211, whilst keeping maximum possible commonality with the SM1A. It had the same module/enclosure size, enabling replacement of the SM1A during a normal refit. The components altered were:-

- LP compressor: Inlet diameter increased and first three stages lengthened to increase mass flow. RPM increased.

- Turbines: New HP blades, NGVs and discs, and honeycomb tip seals as used in the Tay, enabling 150° higher TET at the same metal temperatures. Similar changes to LP.

- Power turbine: Cooled double skin inlet duct to cope with higher gas generator exit temperature, and changed nozzle materials.

The first run of the SM1C was in 1987, followed by endurance and cyclic running at Ansty, and then a test at Pyestock to demonstrate the compatibility of the SM1C gas generator and SM1A power turbine. HMS *Brave* went to sea with a complete SM1C unit in 1990.

Figure 77
HMS Brave featuring SM1C power units

Spey ICR (inter-cooled, regenerative)

In the mid 1980s, working on a US Navy-sponsored study, Rolls-Royce produced a design for an inter-cooled, regenerative version of the SM1C. This was a return to the origins of gas turbine marine propulsion. Rolls-Royce had lead the world in 1953 with the introduction of the RM60 unit in the Royal Navy s HMS *Grey Goose*. The RM60 was an inter-cooled, regenerative engine which had also excited the interest of the US Navy but it was before its time and no further use was made of the idea until the Spey ICR over 30 years later.

Marine propulsion poses particular problems for the gas turbine, one of the most significant being that cruise power is only about 30% of maximum power and at this power the SFC is about 40-50% worse than at maximum power. This problem had hitherto been overcome by using several engines and shutting some down for cruise, but this was complex, heavy, costly and space-consuming. The answer is to transfer some of the waste energy in the exhaust gas back into the air before it enters the combustion process using a heat exchanger. This process can be made even more efficient if the air is cooled during the compression process, hence the inter-cooling, and a variable area power turbine is used.

Considerable design work took place on the ICR Spey in conjunction with Allison, particularly following a RFP from the US Navy in 1989, in which Rolls-Royce were in competition with PW and the formidable opponent GE who powered virtually all the USN s gas turbine-powered ships. In 1990 the USN suddenly increased the maximum power requirement from 26,400 hp to 30,000 hp. It was possible to achieve this power with the ICR Spey but only with considerable compressor and turbine changes, whereas use of existing –535 IP and HP compressors and –524 HP and –535 IP turbines would help reduce development costs and improve service reliability. Thus the WR21 engine was born, which eventually won the competition and is now being successfully developed to meet not only the US Navy s specification but also the Royal Navy s and the French Navy s as well. There is no doubt that the work done on the ICR Spey, even though it was never built, contributed considerably to the success of the WR21.

Figure 78
The WR21 Marine Gas Turbine

Chapter ten: Afterthoughts

Choice of optimum bypass ratio – did we choose the right bypass ratio?

Over the years – and with the benefit of 100% hindsight – various so-called experts have claimed that Rolls-Royce got the optimum bypass calculations for the Spey wrong and the figure should have been much higher. Choice of bypass ratio for an engine produces more emotion than almost any other aspect of the engine design – yet it is not a fundamental parameter. The fundamental parameter is specific thrust or thrust/airflow which, in turn, decides the jet velocity (Vj) of the engine and hence for a given aircraft velocity the propulsive efficiency;

propulsive efficiency = 2 x aircraft velocity/jet velocity + aircraft velocity: = 2Va/(Va + Vj), but net thrust = airflow x (Vj - Va)

To achieve a high propulsive efficiency Vj must be close to Va – but for 100% efficiency Vj=Va and no thrust is produced! Thus a compromise is necessary to ensure a high propulsive efficiency as well as a satisfactory thrust/weight ratio. In choosing the optimum specific thrust for an engine, bypass ratio is used as a variant in any parametric study only because it is a more convenient input into the exercise – not because it is a fundamental parameter, like pressure ratio or turbine entry temperature.

The objective in trying to achieve a high propulsive efficiency at cruise is, of course, to lower the fuel consumption and hence the aircraft operating costs. But an airline's profits depend not only on minimising costs but also on maximising revenue and here engine weight plays a significant role. Any weight added to an engine means, in the ultimate, that the aircraft carries less passengers and revenue is thus reduced. Revenue acts directly on the 'bottom line' whereas fuel costs, although it is the biggest single bill, are a fraction of an airline's total costs – typically less than a sixth. Thus in deciding the optimum cycle for an engine one has to evaluate those features which improve the fuel consumption but do not impair revenue earning capability by increasing engine weight.

The engine optimisation process has to consider all of the following engine parameters: overall pressure ratio, turbine entry temperature, component efficiencies, total powerplant weight, nacelle drag, aircraft velocity, cost and many other items covering reliability, ease of operation and environmental friendliness.

The Spey cycle optimisation was extremely thorough, taking most of the above parameters into account; noise was considered but it was considered that emissions would be minimised by the combustion department whatever the cycle. The exercise was certainly the most detailed of any that had been carried out by either engine manufacturer, aircraft constructor or airline up to then since it covered revenue earning capability as well as operating costs. As mentioned in the main text, the curve of revenue earning capability vs bypass ratio was fairly flat and the optimum was chosen at 1.0 because it limited exposure to some of the unknown design problems. The primary parameter, which prevented choice of a higher bypass ratio, was powerplant weight followed by nacelle drag.

The powerplant weight is, of course, mainly made up of the individual component weights plus the nacelle weight. Component weights are largely influenced by aerodynamic parameters, such as mach number and aerodynamic loading, and by disc stressing rules. At that time, Rolls-Royce – like all the engine manufacturers – was striving to achieve the most efficient engines by designing the most aerodynamically efficient compressors and turbines, and various edicts had been laid down restricting the mach numbers and loadings to be used; in this they were very successful. The Spey components were the most efficient of any Rolls-Royce engine for the next 20 years – more efficient than those of the RB211 at launch; in fact, they were not bettered until the late '70s when large powerful computers enabled the use of CFD. However, this high efficiency came at a cost in terms of engine weight, which then influenced the choice of 'optimum' bypass ratio – it was lower than we would think of today but, given the components used, not by much. A glance at Table 1 shows that the cycle of the JT8D is very similar to that of the Spey, so if Rolls-Royce got it wrong so did P&W.

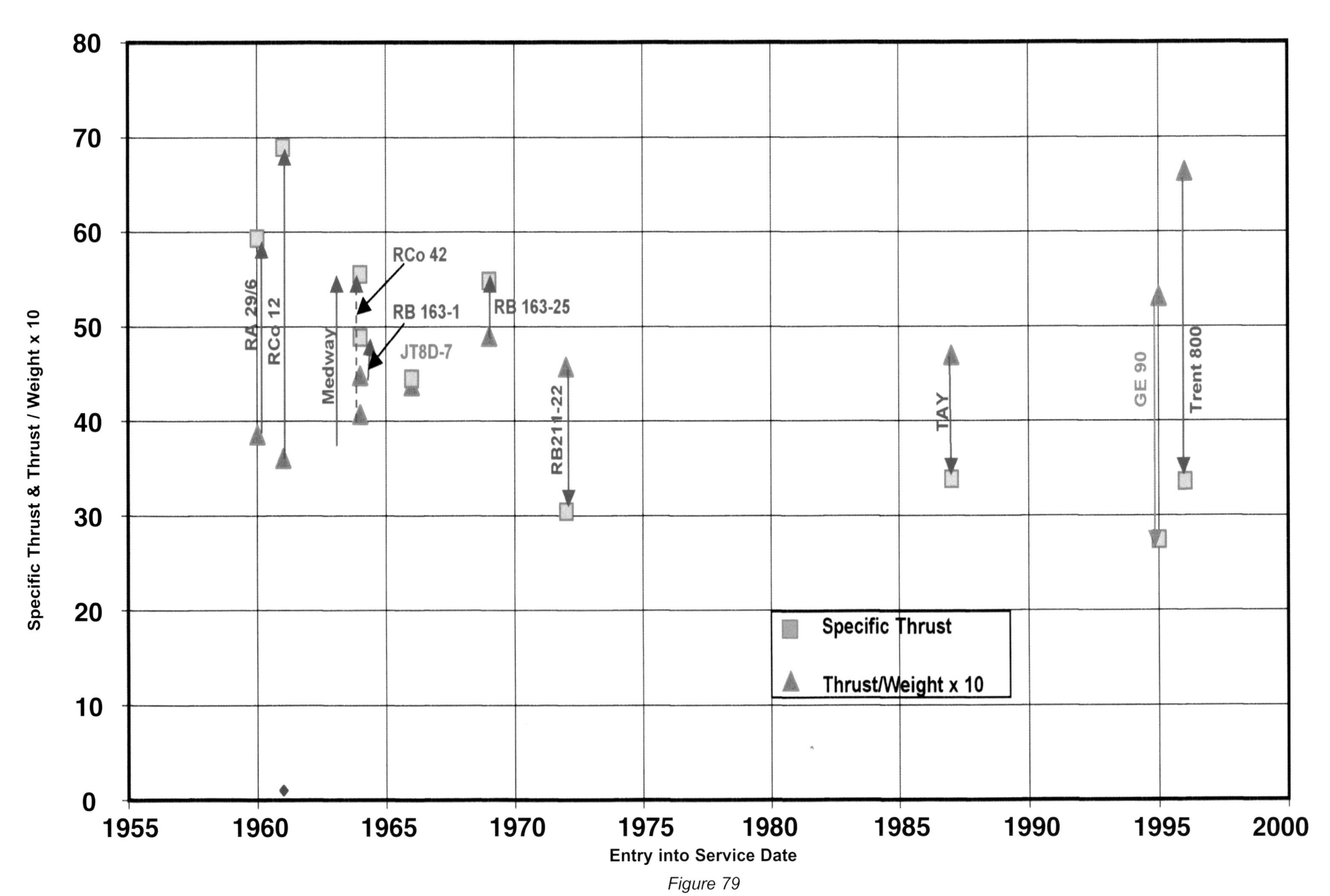

Specific Thrust & Thrust / Weight x 10
Entry into Service Date
RA 29/6
RCo 12
Medway
RCo 42
RB 163-1
JT8D-7
RB 163-25
RB211-22
TAY
GE 90
Trent 800
Specific Thrust
Thrust/Weight x 10
0
10
20
30
40
50
60
70
80
1955
1960
1965
1970
1975
1980
1985
1990
1995
2000

Figure 79

Figure 78 illustrates the dilemma facing the cycle designer over the years. Its shows specific thrust (orange squares) which needs to be low to give good propulsive efficiency and hence low sfc, and thrust/weight ratio which needs to be high to improve revenue earning capability (the plot uses thrust/weight ratio times 10 as this enables the same scale to be used for both parameters). In the early '60s, specific thrust was high but thrust/weight ratio was low – although the Conway was a bypass engine, its bypass ratio was so low that its specific thrust was very high. With the advent in the mid '60s of engines such as the Spey and JT8D, with their bypass ratios around 1, specific thrust improved but not at the expense of thrust/weight ratio which was good for the time. The 163-25, like almost all derivatives, improved thrust/weight at the expense specific thrust – however, sfc was maintained through increased pressure ratio.

A consequence of the choice of a bypass ratio of 1.0 for the civil Spey was that it was around the ideal for a military engine, though this was pure accident – it was not considered at all at the time the choice was made. However, since efforts were being made to install the engine in both an un-reheated version in the Buccaneer and a reheated version (AR 963R) in the TFX, if the bypass ratio had been unsuitable no doubt some changes would have been made for these installations. As it was, the Spey succeeded very well in both these and other military applications. In fact, slightly more military engines were sold than civil. A higher bypass ratio would have resulted in a larger frontal area for a given thrust thus causing higher drag, particularly at high subsonic speeds; a lower bypass ratio would have resulted in a high fuel consumption thus reducing mission radius. Since that date engines specifically designed for military operation, such as the Adour, RB199, F110, F404 or EJ200 have had a bypass ratio of 1.0 or lower.

Figure 78 also shows the effect of the large jump in bypass ratio in the early '70s on specific thrust, but with no corresponding reduction in thrust/weight ratio. The RB211 cycle was enabled by the dramatic increase in aerodynamic loading and mach numbers of the compressors and turbines compared to the Spey, resulting in lower component weight. This also helped to reduce engine cost by reducing blade numbers and the lower jet velocity meant both significantly lower sfc and lower noise.

Since the advent of the high bypass ratio engines, 1970 on, the optimum specific thrust has remained largely constant. The specific thrust of the Trent 800 and even the Tay is about the same as that of the RB211-22B though the bypass ratio has gone up from around 4 or beyond 6 and that is still limited by powerplant weight and the aircraft duty – twin-engined, medium range. The GE90 has a lower specific thrust by virtue of its extreme bypass ratio, about 9, but that is at a penalty of a much worse thrust/weight ratio. Some would argue that is why it has not sold as well as the Trent 800 – it has a slightly better fuel consumption than the Trent 800 but its revenue earning capability is worse. The Trent 500 has a bypass ratio of almost 8 but it is installed in a four-engined, long range aircraft, which makes a difference to the choice of optimum specific thrust. The Trent 900, the engine for the super-large Airbus A380, has a bypass ratio of over 8 but is slightly higher than optimum for noise reasons.

In all of the debate about 'optimum' bypass ratio it is worth noting that the highest propulsive efficiency of any engine in commercial service was that of the Olympus in the Concorde – and that has a bypass ratio of exactly zero!

Some frequently asked questions

The question has often been posed, 'Why did the AR963 not succeed?'. In fact, it did win the competition to power the 727 in the eyes of Boeing and United Airlines, both of whom had traditionally chosen P&W engines. The fact that Eastern, and in particular Eddie Rickenbacker, chose the P&W engine had nothing to do with the choice of cycle for the Spey since a glance at the table in Figure 78 shows that the cycle of the Spey RB163-1, and hence the AR963, was fairly close to that of the JT8D. As Jim Knott, Allison engineering director, said, *"Rickenbacker chose the JT8 not for technical reasons, but because he knew the P&W management and had only recently become acquainted with the Allison and Rolls people".*

It is probably more relevant to ask why, following the rejection of the AR963 by Eastern and the selection of the JT8D, Rolls-Royce and Allison did not continue to pursue the market possibilities for the AR963 with other airlines. They had, after all, succeeded in convincing at least some of Boeing and United Airlines of the merits of the engine. The answer will probably never be known, but it is possible to offer some reasonable possibilities:-

1. In spite of the views of the Advanced Project Office under Jack Steiner, not everyone in Boeing was convinced of the merits of the AR963 or the Rolls-Royce/Allison partnership. In addition, it is unconceivable that Boeing would have launched a new aircraft without having a P&W engine on it. They had always launched a new aircraft with a P&W engine (the –535C on the 757 was the first to break that tradition more than 20 years later), indeed they had rarely had anything other than a P&W engine on their civil aircraft and although the Conway had got on to the 707 that was after the launch with the PW JT3 to be followed by the PW JT4 and then the highly successful JT3D; it is worth noting that Boeing, P&W and United Airlines all once formed part of the same company. In the event, it is highly likely that Rolls-Royce and Allison knew the JT8D would be on the aircraft anyway and did not wish to fight P&W head-to-head in P&W's own backyard.

2. The AR963 in its military form was in a competition to power the swing-wing TFX aircraft, which later became the F111. When P&W won this contest with the TF30, a reheated version of the JT8D, the game was over.

Another question which can be asked is, 'What would have happened had not BEA reduced the size of the DH121?'. Looking at it from the engine point of view, there is no doubt that the larger aircraft would probably have been more popular, but it would have been a direct competitor to the Boeing 727, and the BAC One-Eleven, since it used the same engines as the Trident, would have competed head on with the Boeing 737 and DC9. Sud Aviation were the real losers on the Caravelle as they really did want the Medway; however, the growth version of the aircraft was instead powered by the JT8D.

In the event, the Spey at its size was able to compete in the marketplace very well. The BAC One-Eleven was a very successful aircraft, which sold in good numbers throughout the world. In particular, it penetrated the US market, which not only gave Rolls-Royce valuable operating experience but introduced them to some of the major airlines. This was later to prove of advantage in the campaigns for the RB211 and even the –535 in the Boeing 757. It is also probable that the presence of the civil Spey in the US influenced the decision on the TF41.

The Trident was not as successful as an export as the BAC One-Eleven though it did get into the Middle East and Chinese market. Having been designed to some very exacting and narrow requirements on the part of BEA, the early versions tended to lack operational flexibility and by the time that had been corrected the Boeing 727 was around and, because it was of a size that the market wanted it, it took the lion's share. It is quite likely that had the Trident been of a larger size it would have had a larger market share as it preceded the 727 by about two years, but its initial lack of flexibility would have hampered it against the Boeing 727.

As far as the Spey is concerned, the answer to the question of whether it would have sold more widely had it been larger is probably no. What is certain is that if the Spey (or Medway) had been competing directly with the JT8D it would not have been as profitable. It is also probable that the change in aircraft size was more significant to the British aircraft industry than to the engine industry.

The Spey was of a size that enabled it to fit into existing military aircraft – the Buccaneer and the Phantom – and this gave Rolls-Royce not only a useful sales opportunity but also valuable experience. Had it been bigger that market would probably not have been possible. It also enabled it to 'grow' into the TF41 for the A7, which on its own sold about half the total of all Spey military sales.

Probably the most potent argument is that it was produced in greater quantities than any other Rolls-Royce engine, apart from the Avon if its licensed production is included.

The final irony is, of course, that, due the Spey's smaller core size, it was possible to re-fan it – and hence the Tay was born – keeping the Spey alive even to the present day. This succeeded on the Fokker 100 and the Gulfstream IV, particularly in America. In a delicious twist of irony, it ousted the JT8s from the United Parcels Boeing 727s, which were successfully re-engined with Tay.

Personal impressions

What was it like working in Engineering in the 1950s and 1960s? Plenty of hard work, but we had a lot of fun.

A large number of new engine projects were launched, many of them reaching the test bed, but only a few went into production. There was a man on the end of a telephone seemingly employed to give out new RB numbers (Larry Marks). Things could be done easily and informally, even rather hair-brained ideas, which often came to nothing.

Young graduates coming for an interview at Elton Road in 1952 might have had second thoughts when they saw the uninspiring array of cars outside the executive entrance. The best was a Maigret-type Citroen – not much better than our own MGs, Wolseley Hornets, Rileys and Alvis. Never mind, eventual places on the board were almost certain. One graduate who spent time on the phone to his stockbroker could not wait. He was lured to Boodles to be interviewed for a directorship of the bogus Titan Lime Phosphate Company in Sweden; a wasted journey.

We worked long irregular hours without overtime or standby pay, or free shoes. The first runs of new engines always seemed to take place on New Year's Eve before the clock struck 12. Compressor surge or sparks coming out of the back prevented the engine getting much above idle, but everyone wanted to know if thrust and SFC guarantees were met. Some of us had a dance to go to afterwards, so we turned up in a dinner jacket protected with an 'analysis' mac. Then back to work at 8.00am on New Year's Day!

Then came the thrill when the £1000 salary barrier was broken, and a move to superior digs.

Light relief came with some very clever practical jokes. The highlights were the memorable Spey 'Festivals of Drama' at the Bridge Inn at Shelton Lock, with entertainment provided by Experimental Test Bed, Development and Performance departments.

On one such evening John Hodson and Peter King performed a ballet. At a Spey Operators' Conference delegates squabbled over their Kansas City, Baghdad, Ndola and Naval ratings. The Aer Lingus representative asked for a special rating to get out of Lourdes on a hot day, and was offered holy water injection. The chairman fought back with, *"You sit there with 23 aircraft, which is just under 2½ apiece – it wouldn't be so bad if they were all the same, but there are 17 different ratings, which is .74 per aircraft".*

Then a somnolent test bed control room scene opened with the introduction 'bathed in the warm glow of the oil pressure warning light'.

Another meeting enthused over a proposal for a lead fishing float. Deep in thought, Arthur Hare walked around unaware that his foot was wedged in a waste paper basket. Vivid yellow braces were sported by the Chairman Furnace Beltit, but he was out-twanged by the real Ernest Eltis sitting on the front row.

Abbreviations

ACU	acceleration control unit
ASI	air speed indicator
ATF	Altitude Test Facility
BLC	boundary layer control
BPD	bypass duct
BV	bleed valve
CASC	combined acceleration speed control
F	fuel
FAR	fuel air ratio
FR	reheat full
gph	gallons per hour
H/U^2	turbine work factor
HP	high pressure
IGV	inlet guide vane
IP	intermediate pressure
kts	knots
LP	low pressure
Mn	Mach number
NH	high pressure speed
NL	low pressure speed
OPR	overall pressure ratio
P_2	LP compressor outlet total pressure
P3	HP compressor outlet total pressure
P_3/P_2	HP compressor pressure ratio
P_7	jet pipe total pressure
P_7/P_1	ratio of P_7 to inlet total pressure
P_7/P_0	ratio of P_7 to ambient static pressure
Pcold	inlet total pressure of cold stream to reheat system
Phot	inlet total pressure of hot stream to reheat system
PRCU	pressure ration control unit
SFC	specific fuel consumption
SL ISA	sea level international standard atmosphere
SQ	square
STOL	short takeoff or landing
T/O	takeoff
T_2	LP compressor outlet total temperature
T4/T6	overall turbine temperature ratio
TET	turbine entry temperature
TGT	turbine gas temperature
Va/U	ratio of turbine axial velocity to blade speed
VIGV	variable inlet guide vane
Vmax	maximum velocity
VMO	variable metering orifice

Overheat

During flight testing of the BAC One-Eleven it was found that under certain approach-to-stall conditions the wash from the aircraft wings could result in severe inlet distortion, which led to deep engine HP compressor stall. This was a recurring problem encountered on all Speys (and also some other engines), when the TGT rose at a fast rate inconsistent with the rate of change of NH, P3, etc. The engine stayed in a stalled condition, and was not able to be rescued by the Top Temperature Control, which had insufficient authority to cut the fuel adequately. The result was a severe 'overheat', which often resulted in burnt out turbine blades. Incidents in service were initiated during fixed throttle running, deceleration and acceleration but they could only be consistently reproduced on the test bed by soaking the engine at maximum rating, rapidly decelerating and then slam opening the throttle at low NH, ie, a hot reslam or 'Bodie'.

During 1967-73 a lot of work was done, culminating in a series of Ministry-funded tests on a Phantom Spey using a computerised control system, which reproduced the characteristics of the CASC ACU, but with a much larger range of adjustment. Reslams were performed from various NH with the fuel increased until either surge or overheat resulted. Reslams from 65% NH and below always resulted in overheat, above 71% in surge and an overlap in between. Figure 79 shows a reslam from 65% NH (1) with the normal ACU, then with 60% more fuel than the ACU (2), and finally 61% more (3). The second resulted in a faster re-acceleration than the first, but the slightly higher fuel in the third put the engine into an overheat, where it remained. This showed that overheat was a cliff edge phenomenon without previous warning, and the P3 oscillation of about ½ HP order indicated that the HP compressor was stalled. The HP compressor graph (Figure 80) shows that the transient working line turned left (red) instead of right (green), finishing up at a point on a secondary characteristic with about 30% less airflow than the primary one. The resulting increase of fuel/air ratio caused the overheat. By dipping the fuel sufficiently, the operating point moved to a stall drop-out point, and then accelerated normally on the primary characteristics (orange).

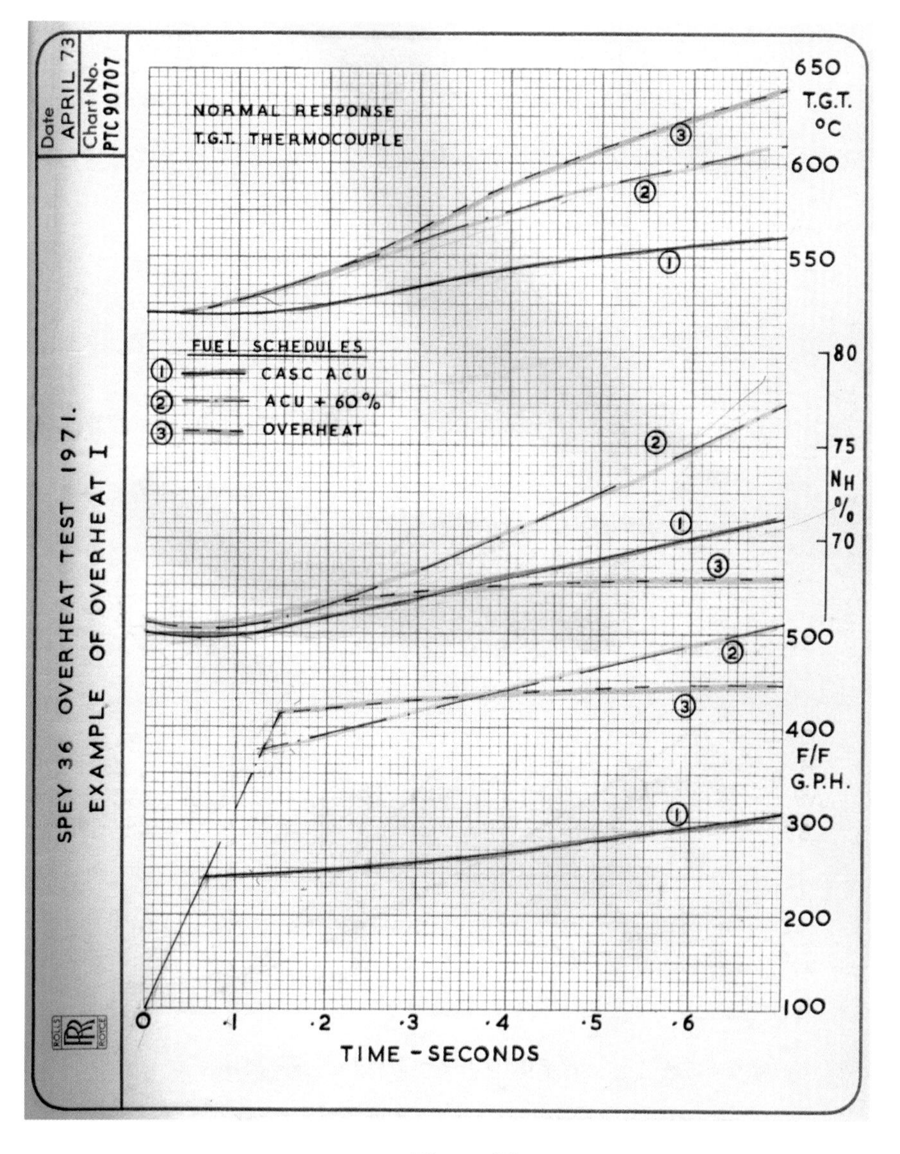

Figure 80

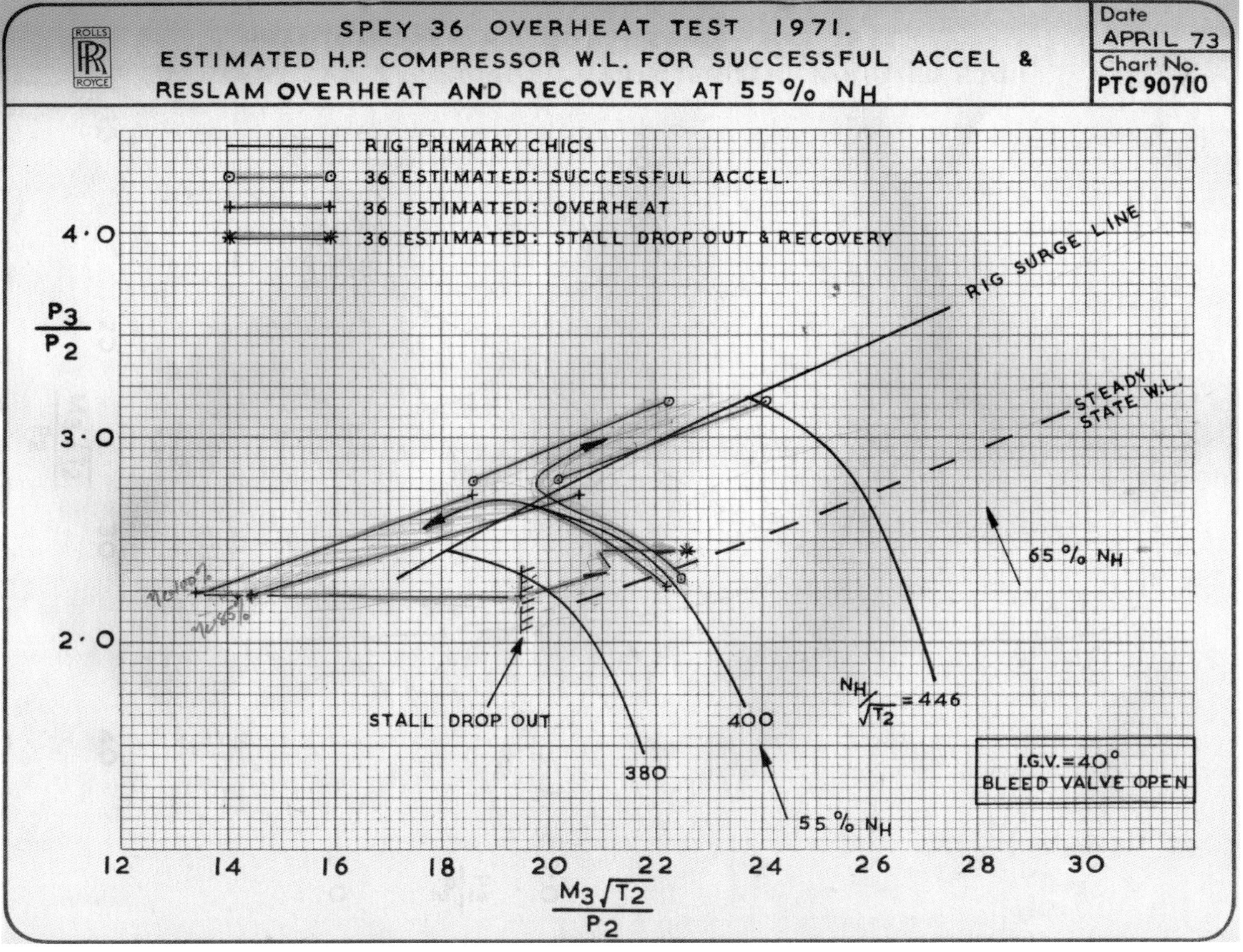

Figure 81
Spey 36 Overheat Test 1971

Up to then, the HP compressor rig was run with a much larger volume between compressor exit and throttle than the engine combustion chamber. For a special test it was reduced to near the engine value, and a set of secondary characteristics were readily obtained similar to those measured on the engine. Stage characteristics showed that all the stages were stalled. Like the engine, the compressor could only be forced there via the primary surge line. The only way the overheat could be reproduced on the engine computer programme was with these measured secondary characteristics.

The larger the exit volume, and the flatter the trajectory, reduced the likelihood of intersecting the secondary characteristics. A larger combustion chamber being out of the question, the only way to avoid overheat was to increase steady state and transient surge margins. As the engine behaved normally up to surge, there was no way of forecasting overheat and preventing it. A cure was needed to get out of the overheat.

When an engine surges, it recovers after the cause has been removed, possibly an aircraft manoeuvre. In overheat it needs a big dip in fuel flow to increase the HP compressor flow beyond the stall drop out. The dipper was triggered by a TGT signal about 15º higher than the TTC, which quickly reduced the fuel flow to about idle by spilling main fuel into the LP. After five seconds the fuel flow was returned to CASC control. The dosage was repeated if necessary. The dipper was manually armed, separately from the TTC, so that it could be disarmed during takeoff for example. The level of TGT signal and level of dipping depended on the engine installation. On some aircraft the dipper was not fitted.

Production engine mark numbers and brief description of differences

As mentioned in the Introduction, the Spey had more different types of application than any other Rolls-Royce engine and, as a consequence, had a plethora of Mark numbers. The table overleaf gives the major mark numbers for civil variants and a brief description of the major differences. It is by no means the complete listing. To gain an impression of the vast numbers, the chart 'Civil Spey Variants' is included although, as it is dated February 1968, it is by no means the total list. This was produced to try and keep track of some of them. At a later date, all Service Engineers and many others were issued with the booklet TSD 4016, which listed the mark numbers of all Rolls-Royce aircraft engines. The total number for the Spey is 94, but that does not include the Tay or the TF41, which was in the US Military system.

Production quantities of the Spey

The following table lists the total quantities of Spey engines produced, including core kits for those assembled outside Rolls-Royce such as the TF41 and the Industrial and Marine engines, which are still in production. Included at the end are the figures for the Tay engine, which of course also includes the Spey core; this is also still in production for the Gulfstream. The grand total is just under 8000 and with production still ongoing on the marine and Tay versions it is possible that a figure of over 8000 can be reached. This comfortably exceeds the Dart at 7100 and is about on par with the Avon with 8129 Rolls-Royce-manufactured engines, though 2639 license-built Avons were also produced by the likes of Standard, Napier, etc.

Aircraft/installation	*Engines/kits produced*	*Comments*
Civil types		
Trident 1, 2, 3	495	Mks 505, 06, 10, 11, 12 and variants of these
BAC One-Eleven	677	Mks 505, 06, 10, 11, 12 and variants of these
BAC One-Eleven (Romania)	24	Mks 511, 512
F28 – Fellowship	613	Mk 555
Gulfstream II and III	959	Mk 511-8
Civil total	2768	
Military types		
Phantom F4K and F4M	542	With reheat. Mks 202, 03, 04
Buccaneer	404	Non-reheat. Mk 101
Nimrod	248	Mks 250, 805, 06
AMX	267	Kits supplied to Italy and Brazil for assembly Mk 807
LTV A7U	1417	Mk 202 parts (~50%) supplied to Allison for assembly
Military total	2878	
Industrial & Marine		
Marine	148	Ongoing production
Industrial	26	
Ind & Marine total	174	
Grand total Spey	5820	
Total Tay production	1991	Gulfstream II and III: 1077 ongoing Fokker 100: 738 Dee Howard (B727 and BAC One-Eleven re-engine): 176
TOTAL SPEY AND TAY	**7811**	

Aircraft and installations of civil, military, and industrial and marine versions of the Spey

Tables 2 and 3 give details of the various civil and military aircraft and gas turbine installations in which the various versions of the Spey are installed. The essence of these tables is to give a flavour of the many different roles in which the Spey took part. It is not intended to be the definitive description of each of these aircraft or installations as the data has come from many different sources, but it has been cross-checked as far as possible. The tables show that, apart from helicopters, there is hardly a gas turbine role in which the Spey engine has not been involved in some form or another.

Takeoff thrust ratings

One of the notable features of the Spey family was the plethora of different takeoff thrust ratings both for the basic four-stage LP version and the uprated five-stage version (read the sixth paragraph in the section on Personal Impressions). The major ones for the Mark 505/6 and for the 510/11/12 on Figure 81. The ratings shown are by no means the total; one of the more extreme was the Lhasa rating, which sounded very painful. The main reason for large numbers of different ratings was that the aircraft were basically under-powered. The original specification for the Trident defined by BEA called for an aircraft to carry out a short-range flight from London to Paris at 600 mph at minimum cost and in order to fulfil these requirements de Havilland designed a 'point' aircraft – that is one which would carry out the design mission and not much else. When BEA and other airlines tried to operate over longer ranges the MTOW (Maximum TakeOff Weight) had to be increased to carry the extra fuel and more thrust was required.

It is also true that there was much more room for growth in the RB141 Medway than in the RB163 Spey – the former grew relatively easily up to 14,000 lb (16% growth) with only minor changes to hardware whereas the latter needed an additional compression stage and a reduction in bypass ratio to achieve 14% growth.

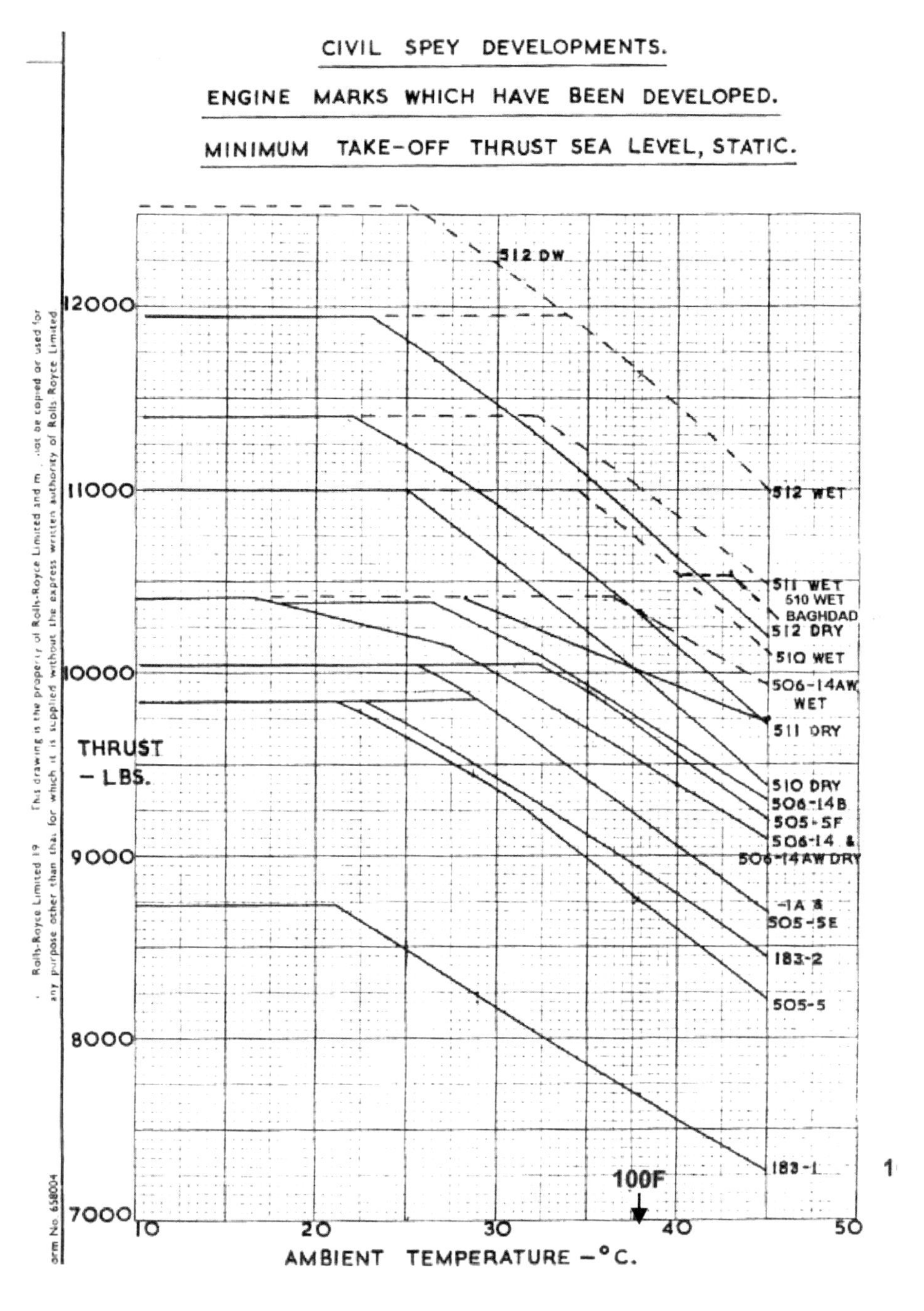

Figure 82
Civil Spey Developments

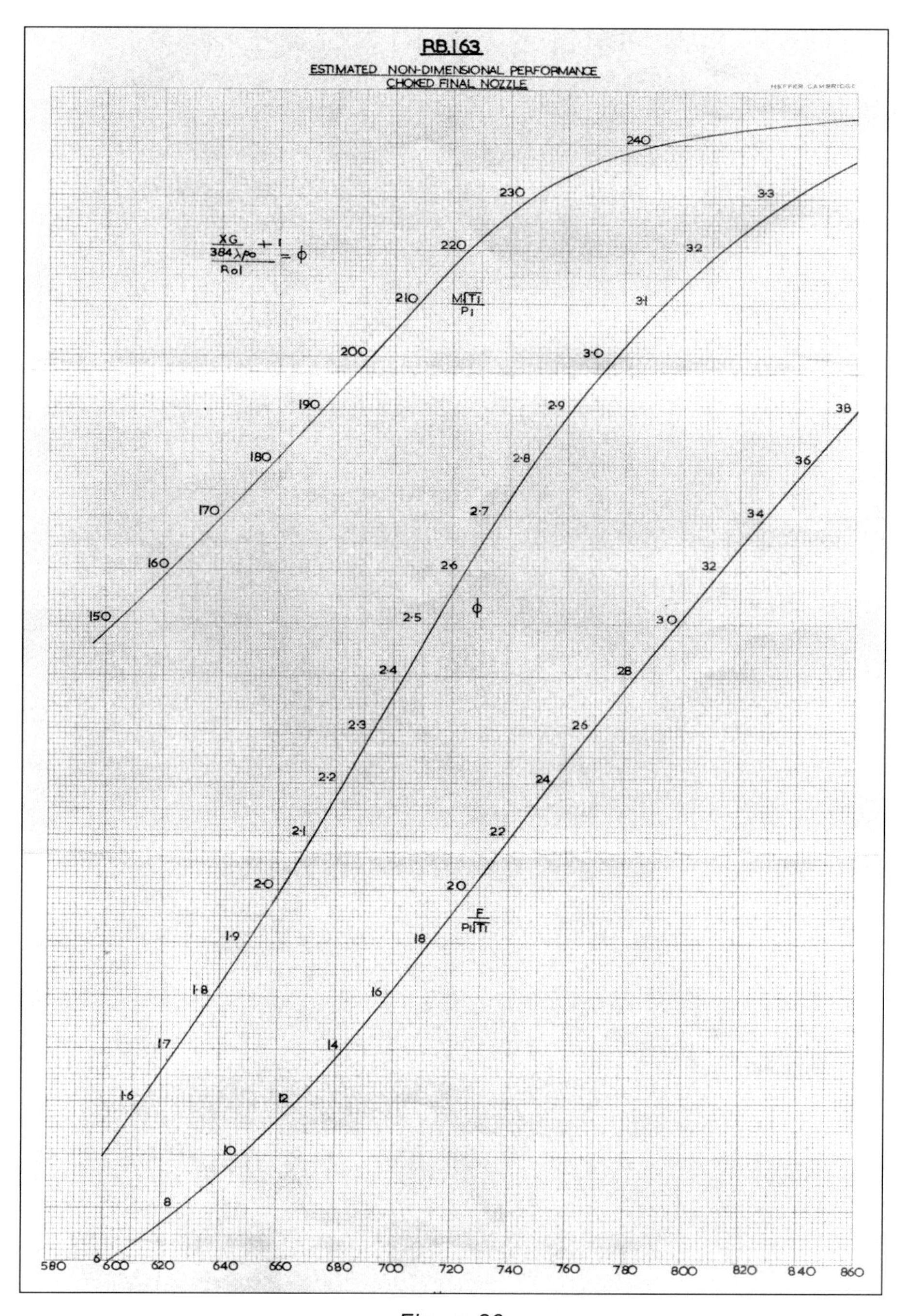

Figure 83

In addition, the arrival of the jet age introduced takeoff thrust as a major problem. Straight-winged propeller aircraft had relatively plenty of thrust available for takeoff, especially with the variable pitch propeller; the limiting thrust condition tended to be climb and cruise. The high-speed swept-wing jet aircraft introduced the need for much longer runways. Most of the large airports had the money to pay for these but in some cases smaller airports had not yet found the cash to pay for the extensions so special airport ratings were required. At other airports there were obstructions, such as tall buildings, high ground or even tall trees, which caused a hazard during climb-out and demanded a special rating. It was all part of the learning process as jet travel spread to all parts of the globe.

Another factor in the multiplicity of ratings was the flat rating ambient temperature. The Spey was flat-rated, ie, produced a fixed level of thrust, up to an ambient temperature of just over 20°C. This was adequate for air travel between the major industrial centres in Europe, which were largely towards the north of the continent. When the BAC One-Eleven was sold to airlines such as American and Braniff, both of which were based in Texas, the need for flat rating to much higher ambient temperatures was essential, hence the advent of boosting by water injection. Most of the 'water' ratings are flat rated well beyond today's norm of 30°C.

Performance

All the initial calculations for the Medway and Spey performance were made on slide-rules as computers had not yet arrived on the scene. The design point calculations were relatively straightforward, but the off-design performance calculations were long and complex as they involved at least two major iterations, one inside the other, as well as many minor iterations. This inevitably introduced scatter and the possibility of errors – much time was spent searching out the causes of 'kinks' in the curves produced. The Project Assessment Office used 12-inch slide-rules for speed of working as they were constantly evaluating various preliminary designs. When the configuration was more or less settled, the performance responsibility was passed to the main Performance Office who looked after all the performance tasks such as issuing brochures, development engine performance analysis and flight-test performance evaluation. Here the norm was the 21-inch slide-rule with optional magnifying glass on the cursor, or the noisy Freiden mechanical calculators, which initially were rather slow because they could not do square roots – except by using a complicated routine known only to the most experienced operators; they were also the people who knew the calculations that made the machines perform rhythmic tunes.

The basic performance curves produced plotted all the engine parameters in non-dimensional form, see Figure 82. The main non-dimensional parameters were:-

Shaft speed	$N/\sqrt{T_1}$	
Thrust function	$[(Xg/Apo) + 1]/R_{01} =$	
Airflow	$M_1\sqrt{T_1}/P_1$	
Fuel flow	$F/P_1\sqrt{T_1}$	
Pressures	Pn/P_1	
Temperatures	Tn/T_1	

where:
T_1 is inlet total temperature Absolute
N is shaft speed
Xg is gross thrust
A is final nozzle area
R_{01} is the ram ratio
P_1 ambient static pressure
P_1 is intake total pressure Absolute
M_1 is intake total airflow
Pn is a numbered total pressure absolute
Tn is a numbered total temperature absolute
T_1 is intake total temperature Absolute
F is total engine fuel flow

These non-dimensional curves allowed the calculation of engine performance at any altitude and forward speed. To obtain net thrust, one had to calculate the gross thrust using and the above equation, and then subtract the intake momentum drag calculated using the airflow equation and the aircraft velocity; SFC was obtained by dividing the fuel flow by net thrust.

The use of non-dimensional curves implied that the specific heats used in the basic calculations did not vary with changes in intake total temperature; the same applied to the effect of Reynolds number on component efficiencies. At a later stage, correction curves were supplied to cater for these minor variations, but in the initial days of the Medway and Spey these would have implied an unwarranted accuracy.

As mentioned above, this was all before the advent of computers. When they did arrive in the late '50s they were vast, cumbersome and very slow machines full of glowing electronic valves – not a transistor or a chip in sight. Because they did not involve iterations, the programmes to calculate performance at design point, or to obtain dimensional performance from non-dimensionals, were relatively simple and well within the capability of the device and proved a boon. But the calculation of off-design performance was an entirely different matter. The programme to do the calculations took quite a time to develop and, due to capacity limitations of the machine, had to be divided into three separate parts – one to prepare the input data, one to carry out the iterations and the final one to formulate and print out the results. The middle programme frequently had to be coaxed through the iteration process and took 45 minutes or even more for a single point! A good operator on a slide-rule could do it faster but the computer reduced the scatter (it also worked at night) and, provided input was good, gave consistent and reasonably accurate answers. What is more, it used fully variable specific heats rather than mean specific heats – that is the gas properties were correct for the temperature and pressure at each point in the calculation.

When the Medway was finally put on to the computer, relative to the hand calculations the thrust increased by nearly 1000 lb and the sfc improved by 3%! One hoped the engine knew all about fully variable specific heats which, of course, it should, but Harry Pearson quite rightly pointed out that previous reconciliation of calculated performance and rig test efficiencies with actual performance was based on mean specific heats. This obviously was part of the reason for the large performance miss on the initial engines and contributed to the need for 'Hartley Factors' in later performance estimates.

Later in the life of the Spey the capability of the computers and programmes used were vastly increased and through various stages, including brochure reading programmes, non-dimensionals were discarded, and the current state of art reached where a performance deck is supplied, which fully calculates the performance using component characteristics, including the correct specific heats and Reynolds number effects, at any point in the flight envelope.

Post exit thrust

A glance at the headings for any early RB163 dimensional performance chart reveals the phrase 'Includes Post-Exit Thrust'. These few words referred to a factor of bizarre origins and dubious merit. When the initial performance of the RB141 was being evaluated it was decided to include the performance benefits and losses of a convergent/divergent nozzle. This is a type of final nozzle in which the exit area is slightly higher than the throat area. Its aim is to gain the benefits of full expansion of the jet at cruise where, due to the forward speed of the aircraft, the pressure ratio across the nozzle is fairly high; the sfc benefit of the con-di nozzle was of the order of 1.5%. However, at takeoff, the nozzle was over-expanded and a thrust loss of about 0.5% was incurred. Initial performance brochures were issued including the benefits and losses of the con-di nozzle and these were included in the aircraft performance and guarantees.

In order to substantiate the con-di nozzle performance, Stan Smith (Tom Frost?) carried out a series of tests in a small wind tunnel at cruise and takeoff conditions on a model representing the whole nacelle and flow through the engine with both convergent and con-di nozzles. The results of the tests confirmed the figures for the con-di nozzle but imagine the surprise when it was found that the convergent nozzle gave exactly the same performance as the con-di nozzle at cruise. Further testing and analysis showed that flow of air round the outside of the nacelle at cruise aircraft speeds effectively formed a convergent-divergent nozzle even though the hardware was a convergent nozzle. Thus there was no

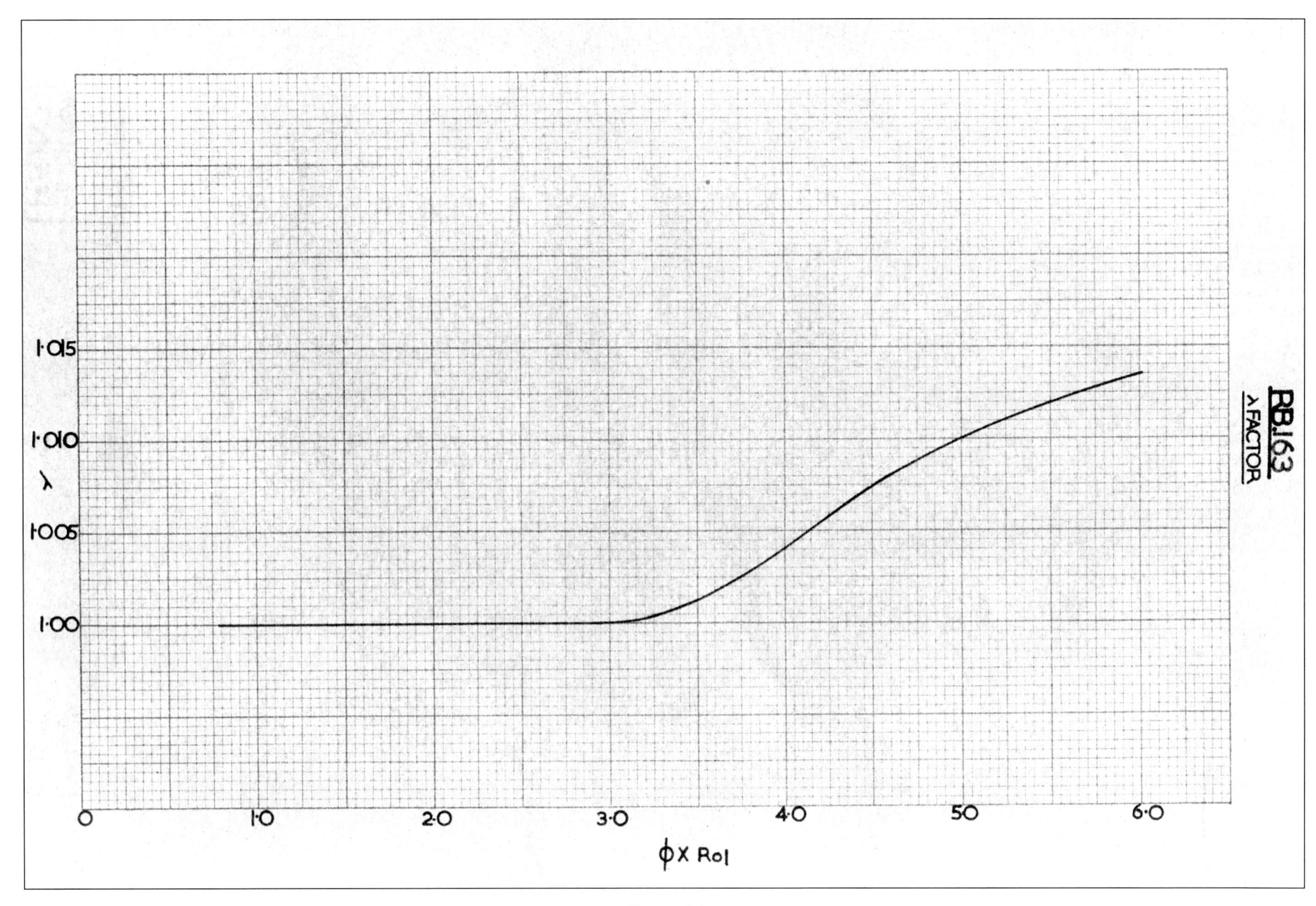

Figure 84

need to provide a con-di nozzle with associated losses at takeoff and additional weight. The problem was how to 'honour' the performance that had already been quoted for the engine with a con-di nozzle.

The solution was to include in the performance brochure a correction factor called 'post exit thrust' (see Figure 87), which exactly mirrored the con-di performance at cruise but with no takeoff losses, accompanied by some 'weasel words' written by Roland Fitzgerald, which tried to justify its inclusion. Naturally there was much heated debate about the correctness of Rolls-Royce claiming the benefits of a con-di nozzle when the engine did not have one, and especially as the benefits were realised by an increase in static pressures on the exterior of the aircraft, such a benefit rightly belonged to the aircraft performance. 'Tommy' Thomas, the de Havilland Chief Performance Engineer, included the factor because he had already agreed the performance guarantees of the aircraft to BEA but later, much to the relief of many, the factor was quietly dropped both by de Havilland and Rolls-Royce.

A period of intense activity

While the whole of the RB141 and 163 programme could be regarded as one of intense activity, the year of 1959 was one in which the future of the family was mapped out as the following summary list of reports and actions shows:-

Decision on need for reduced size DH121 and hence for a smaller engine	Spring 1959
Further bypass ratio optimisation study (resulting in choice of 1.0)	Easter 1959
Preliminary performance estimates and design layout of RB163	May/June 1959
Psn meeting at which go-ahead for the RB163 was given	01 July 1959
Issue of single Design Point info for RB163-1: 9830 lb TO thrust	02 July 1959
Issue of non-dimensional performance for RB163-1 (TET = 1295K @ ISA, 1325K @ 100F)	06 July 1959
Request for increased takeoff thrust by de Havilland; 10,100 lb instead of 9830 lb	14 July 1959
Revised non-dims supplied to de Havilland: 10,100 lb TO thrust	27 Aug 1959
Boeing (Jack Steiner) request for 10-20% more thrust than RB163 for the Boeing 727	22 Sept 1959
Proposals for Boeing 727 including scaled RB163 from AGN and RMF (later AR963)	23 Sept 1959
First visit by technical team (Hur and Clns) to Blackburn for talks on RB168 in NA39	27 Oct 1959
First issue of performance for proposed AR963 and mention of military rating	06 Nov 1959
Preliminary performance data and assessment of flap blowing for NA39 issued	22 Nov 1959
First run of RB141	23? Nov 1959

Table 1: Details of Spey engines compared to other Rolls-Royce engines and competitors' engines

Engine type	Avon Ra 29/3	Avon Ra 29/6	Conway RCo 12 RB80 Mk 508, 9	Conway RCo 42 Mk 540	Medway RB141	Spey RB163-1,2,3 RSp 1,2,2W,3,3W Mk 505/506	Spey RB163-25 RSp 4,5 Mk 511/512	Spey RB183 Mk 555
Engine performance								
Takeoff:								
Thrust lb SLS ISA	10,500	12,600	17,500	20,370	~14,000	9850-10,400	11,400-11,950	9850
Flat rating temp ºC	15	15	15	11		21-36.5 (W)	22-34 (W)	23
Cruise:								
Altitude ft	36,090	36,090	36,090	36,090	36,090	36,090	36,090	36,090
Mach no	0.83	0.83	0.83	0.83	0.83	0.83	0.83	0.83
SFC lb/hr/lb	0.937	0.918	0.875	0.82	0.8	0.775	0.78	0.767
Cycle parameters @ TO ISA:								
Overall pressure ratio	9.3	10.33	15.3	16	16.75	16.5	18.4	16.5
Bypass ratio	0	0	0.3	0.6	0.7	0.96	0.67	0.96
Turbine entry temp ºK				1255	1295	1295	1295	1295
Specific thrust lb/lb/sec	60.7	68.9	59.32	55.5	54.9	48.8	54.8	49
Air mass flow lb/sec	173	183	295	367	255	202	208	202
Dimensions:								
Length in	126	134	135.9	154	133.9	110	114.6	110
Fan/inlet diameter in	38.4	39.4	37.5	42	37.64	34.25	34.25	34.25
Basic engine weight lb	3347	3490	4544	5001	3738	2200	2332	2023
Thrust/weight lb/lb	3.14	3.6	3.85	4.07	3.74	4.48	4.89	4.89
Thrust/weight x 10	31.4	36	38.5	40.7	37.4	44.8	48.9	48.9
Configuration:								
Compressor stages	16-axial	17-axial	7+9	4+3+9	5+11	4+12	5+12	4+12
Combustor type	Can 8	Can 8	Can 10	Can 10	Can 10	Can 10	Can 10	Can 10
Turbine stages	3	3	1+2	1+3	2+2	2+2	2+2	2+2
Pressure ratio/stage compr	1.15	1.15	1.19	1.19	1.19	1.19	1.19	1.19
Applications:	Comet 2,3 Caravelle	Comet 4 Caravelle	B707 DC8	VC10 Super VC10	Dev't only	Trident 1 BAC One-Eleven	Trident 2,3, 3B BAC One-Eleven Gulfstream II, III dH Buffalo HS Nimrod	Fokker F28
Timescale:								
First engine run	1955	1959	1956	1960	1959	1960	1965	1965
Certification	1958	1960	1959	1963		1963	1968	1968
Entry into service	1958	1961	1960	1964		1964	1969	1969

Engine type	Spey RB168-1 RSp 2 Mk 101, 807	Spey RB168-25R RSp 5R Mk 201-4	Spey RB168-62, -66 TF41-A-1, 2	RB211 -22B	Tay Mk 611	P&W JT8D-7	P&W JT8D-15	Trent 884	GE 90 85B
Engine performance									
Takeoff:									
Thrust lb SLS ISA	11,150	12,250 RHU 20,515 RHL	A1 14,500 A2 15,000	42,000	13,850	14,000	15,500	86,910	85,000
Flat rating temp ºC	15		25	28.9 (B)	30	28.9	28.9	30	30
Cruise:									
Altitude ft				35,000	35,000	30,000	30,000	35,000	35,000
Mach no				0.85	0.8	0.8	0.8	0.83	0.83
SFC lb/hr/lb				0.628	0.71	0.796	0.81	0.575	0.56
Cycle parameters @ TO ISA:									
Overall pressure ratio	18	20	20	24.5	15.8	15.8	16.5	38.8	39
Bypass ratio	0.95	0.66	0.76	4.8	3.04	1.1	1.05	5.9	8.6
Turbine entry temp ºK	1400	1425	1425	~1500					
Specific thrust lb/lb/sec	54.7	58.3/97.7	55.2	30.4	33.8	44.44	48.14	33.55	27.4
Air mass flow lb/sec	204	210	258	1380	410	315	322	2590	3100
Dimensions:									
Length in	110	213	102.6	119.4	94.7	123.5	123.5	172	193
Fan/inlet diameter in	34.25	34.25	37.5	84.8	44	40.5	40.5	110	123
Basic engine weight lb	2255	2605	3175	9195	2951	3205	3414	13,100	16,000
Thrust/weight lb/lb	4.94	4.7	4.49	4.57	4.69	4.37	4.54	6.63	5.31
Thrust/weight x 10	49.4	47		45.7	46.9	43.7	45.4	66.3	53.1
Configuration:									
Compressor stages	4+12	5+12	3+2+11	1+7+6	1+3+12	6+7	6+7	1+8+6	1+3, +10
Combustor type	Can 10	Can 10	Can 10	Ann 18	Can 10	Can 9	Can 9	Ann 24	Ann Dual
Turbine stages	2+2	2+2	1+1+3	2+2	1+3	1+3	1+3	1+1+5	2+6
Pressure ratio/stage compr	1.2	1.19	1.19	1.26	1.19	1.24	1.24	1.28	1.3
Applications:	Blackburn Buccaneer	McDonnell Phantom F4K & F4M, Chinese F7	LTV Corsair A7-1, 2	Lockheed Tristar L1011-1 L1011-100	Gulfstream IV Fokker 100 Fokker 70				
Timescale:									
First engine run	1960	1965	1967	1969	1984				
Certification	1965	1967	1968	1972	1986				
Entry into service	1965	1968	1969	1972 1973(B)	1987	1966	1971	1996	1995

Table 2: Details of commercial aircraft fitted with Spey engine

Aircraft name	Trident 1	Trident 1E	Trident 2E	Trident 3B	One-Eleven	One-Eleven	One-Eleven	One-Eleven	One-Eleven
Type (1)	SRP	SRP	S/MRP	SRP	SRP	SRP	SRP	SRP	SRP
Maker	HSA	HSA	HSA	HSA	BAC	BAC	BAC	BAC	BAC
Variant/Series:	1	1E	2E	3B	200	300	400	475	500
Dimensions:									
Span ft	89.8	95	98	98	88.5	88.5	88.5	93.5	93.5
Length ft	114.75	114.75	114.75	131.2	93.5	93.5	93.5	93.5	107
Height ft	27	27	27	28.25	24.5	24.5	24.5	24.5	24.5
Wing area ft2	1358	1446	1461	1493	1003	1003	1003	1003	1031
Weights:									
Maximum takeoff lb	115,000	135,500	143,500	150,000	79,000	88,500	88,500	98,500	104,500
Empty lb	67,500	71,500	73,200	81,778	46,405	48,722	49,587	51,731	53,911
Payload lb	22,000	25,170	21,378	33,772	19,095	22,278	21,413	21,269	27,089
Passengers maximum/normal	103/88	139/88	149/97	180/150	89/65	89/65	89/65	89/65	119/97
Crew	3	3	3	3	2	2	2	2	2
Powerplant:									
Number and type	3 x Spey 1	3 x Spey 25	3 x Spey 25	3 x Spey 25	2 x Spey 1	2 x Spey 25	2 x Spey 25	2 x Spey 25	3 x Spey 25
Mark No	505-5	511-5	512-5W	512-5W	506	511	511	512-14DW	512-14DW
Thrust MTO ISA	9850	11,400	11,930	11,960 1 x RB162-86 5250	10,330	11,400	11,400	12,550	12,550
Performance:									
V cruise kts/Mach (2)	511/0.85	525/0.85	525/0.85	581	541	541	541	541	541
Range @ typical payload nm	1290	1710	2130	1440	760	1240	1240	1620	1285
Production:	24	15	50	26	56	9	70	9	80
Dates:									
First flight	Jan 1964	1 Nov 1964	Jul 1967	Dec 1969	Apr 1965		Jul 1965	Aug 1970	Jun 1967
Entry into service	Mar 1964	Late 1965	Apr 1968	Apr 1971	Apr 1965		Nov 1965	Jul 1971	Feb 1968
Customer airlines: (3)	BEA	Air Ceylon Channel Northeast Iraqi Kuwait CAAC	BEA Cyprus CAAC	BEA	BUA Braniff Mohawk Zambia Aer Lingus Aloha RAAF	Brit Eagle Brit Midland Laker	American Philippine Tarom Cambrian Austral Bavaria etc etc	Faucett Pacific Malawi	BEA Caledonian Court Line Germanair Romania

Notes:
(1) SRP – Short range passenger; MRP – Medium range passenger (2) At operational range (3) Main and early customers

Aircraft name	F28	F28	F28	F28	GII	GII
Type (1)	SRP	SRP	SRP	SRP	Business	Business
Maker	Fokker	Fokker	Fokker	Fokker	Gulfstream	Gulfstream
Variant/Series:	Mk 1000	Mk 2000	Mk 3000	Mk 4000	GII	GII
Dimensions:						
Span ft	77.35	77.35	82.25	82.25	71.75	77.15
Length ft	89.9	97.1	89.9	97.1	79.9	82.9
Height ft	27.8	27.8	27.8	27.8	24.5	24.5
Wing area ft2	822	822			810	935
Weights:						
Maximum takeoff lb	65,000	65,000	70,800	70,800	65,500	68,200
Empty lb	34,120	35,553			37,186	42,000
Payload lb						
Passengers maximum/normal	65/	79/	65/	85/	19 max/12	19 max/12
Crew	2	2	2	2	2	2
Powerplant:						
Number and type	2 x RB183	2 x RB183	2 x RB183	2 x RB183	2 x Spey 25	2 x Spey 25
Mark No	555	555	555-15H	555-15H	511-8	511-8
Thrust MTO ISA	9850	9850	9900	9900	11,400	11,400
Performance:						
V cruise kts/Mach (2)	519	519	421	421	483	511
Range @ typical payload nm	830	470	1400	870	3580	4047
Production:	97	10	15+	42+	256	
Dates:						
First flight	1967	Apr 1968			Oct 1966	Dec 1979
Entry into service	Mar 1968	Oct 1972	Mid 1977	End 1976	Dec 1967	Mid 1980
Customer airlines: (3)	Braathens Ansett Garuda LTU Iberia etc etc	Nigeria	Garuda Argentina Cimber etc etc	Linjeflyg KLM-NLM Altair Air Ivoire etc etc	Numerous	numerous

Table 3: Details of military and research aircraft fitted with the Spey engine

Aircraft name	**Buccaneer**	**Phantom**	**Corsair**	**Nimrod**	**AMX**	**XAC JH-7**	**Buffalo**
Type	Low level strike, carrier	Multi-role, carrier	Low level strike, carrier	Maritime patrol and AEW	Multi-role attack	Multi-role	Research Augmented Wing
Maker	HS (Blackburn)	McDonnell	Ling-Temco-Vought	HS (de Havilland)	IAF (Fiat and Embraer)	Chinese Aircraft Factory	de Havilland Canada
Variant/Series:	S2A-D, 3 and 50	F4K and F4M	A7-D, E	MRA Mk 1, 2, 3		JH-7	Modified C-8A
Dimensions:							
Span ft	44	38.4	38.75	114.8	29.1	42	78.75
Length ft	63.4	57.6	46.1	126.75	44.5	68.9	94.3
Height ft	16.25	16.25	16.1	29.7	15	21.5	28.7
Wing area ft2	Not available	Not available	375	2,121	226	563	865
BLC	Wing and tail 7th HP	Wing 7th and 12th	None	None	None	None	Bypass air to flaps, ailerons. Core air thro' swivelling nozzles
Crew	2	2	1	12	1	2	
Weights:							
Maximum takeoff lb	62,000	58,000	42,000	177,500 normal	27,558	62,776	45,000
Empty lb	~30,000	31,000	19,111	86,000	14,770		32,600
Powerplant:							
Number and type	2 x Spey 2 RB168	2 x Spey RB168-25R	1 x Allison/ RR Spey	4 x Spey RB168-20	1 x Spey RB168-25	Spey RB168-25R	2 x Spey RB163-25 modified
Mark No	101	202/203/204	TF41 A-1, 2	250/251	807	202 (XIAN WS9)	801
Thrust Maximum ISA	11,150	12,250 R/H unlit 20,515 R/H lit	A-1, 14,500 A-2, 15,000	12,140	11,030	12,250 R/H unlit 20,515 R/H lit	11,400 nom
Production (aircraft):	133	162	1068	47	187? Italy 79? Brazil 40? Thai	?	1 for research only
Performance:							
Maximum speed Mach/mph	Mach 0.85 SL	M 1.2 SL M 2.1 Alt	Mach 0.92 SL	575 max 230 patrol	M 0.86 SL	Mach 1.6-1.7 max	184
Maximum ceiling ft	>40,000	~60,000	Not relevant	42,000	43	50,850	Not relevant
Mission radius nm	1,000	~760	~620	~2200	~450	970	300 range
Dates:							
First flight	23 Jan 1962	27 Jun 1966	Autumn 1967	28 Jun 1968	15 May 1984	Late 1988	May 1972
Entry into service	5 Jun 1964	May 1969 (RAF)	1969	2 Oct 1969	1989	1992-3?	mid 1973
Customers:	Royal Navy Royal Air Force South African Air Force	Royal Navy Royal Air Force	US Navy US Air Force Hellenic Air Force	Royal Air Force	Italian Air Force Brazilian AF Thai Air Force	PLA Air Force PLA Naval Air Force	NASA Canadian Govt

Table 4: Spey Marine installations

Ship type	Frigate	Frigate	Frigate	Destroyer	Destroyer	Frigate	Frigate	Frigate	Frigate	Study only
ID number	22, -07	22-11 to 14	23	DD 171	DD 151	DE 229	M	22	M	
Navy	UK	UK	UK	Japan	Japan	Japan	Netherlands	UK	Netherlands	UK
Installation:										
Arrangement	COGOG	COGAG	CODLAG	COGAG	COGAG	CODOG	CODOG	COGAG	CODOG	Any
Spey engine	2 x Spey	2 x Spey	2 x Spey	2 x Spey	4 x Spey	2 x Spey	2 x Spey	2 x Spey	2 x Spey	Spey
Mark	SM1A	SM1A	SM1A	SM1A	SM1A	SM1A	SM1A	SM1C	SM1C	ICR
Power MW	12.75	12.75	12.75	12.75	12.75	12.75	12.75	19.5	19.5	~25
Power BHP	17,090	17,090	17,090	17,090	17,090	17,090	17,090	26,150	26,150	Up to +25%
Other engine	2 x RR Tyne	2 x RR Tyne	Diesels?	2 x RR Olympus		Diesels?	Diesels?	2 x RR Tyne	Diesels	
SFC lb/bhp/hr										
@ maximum power	0.396	0.396	0.396	0.396	0.396	0.396	0.396	0.372	0.372	
@ 30% power	0.525	0.525	0.525	0.525	0.525	0.525	0.525	0.493	0.493	Up to 40% lower sfc
Dimensions:										
Length gas generator ft	10.2	10.2	10.2	10.2	10.2	10.2	10.2	10.2	10.2	
Length module ft	24	24	24	24	24	24	24	24	24	
Height ft	10.2	10.2	10.2	10.2	10.2	10.2	10.2	10.2	10.2	
Width ft	7.5	7.5	7.5	7.5	7.5	7.5	7.5	7.5	7.5	
Weight module, tonnes	25.7	25.7	25.7	25.7	25.7	25.7	25.7	25.7	25.7	

Table 5: Spey Industrial installations

Use or configuration	Gas generator	Mechanical drive	Electrical generation
Performance:			
Exhaust gas power – maximum kW	16,256	14,100	13,500
Exhaust gas power – maximum EGHP	21,800	18,913	
Exhaust gas power – ISO base kW	14,541	12,600	12,100
Exhaust gas power – ISO base EGHP	19,500	16,887	
Thermal efficiency - % maximum con	40.5	35.4	34
Thermal efficiency - % ISO base	40	34.9	33.5
Specific fuel consumption:			
Maximum kj/kWh		10,157	
ISO base kj/kWh		10,306	
Dimensions:			
Length – excluding intake flare ft	8.86		
Maximum width ft	3.74		
Weight lb	3624		
Applications:	Numerous	Numerous	Numerous

Historical Series

1 *Rolls-Royce - the formative years 1906-1939*
Alec Harvey-Bailey, RRHT 2nd edition 1983
2 *The Merlin in perspective - the combat years*
Alec Harvey-Bailey, RRHT 4th edition 1995
3 *Rolls-Royce - the pursuit of excellence*
Alec Harvey-Bailey and Mike Evans, SHRMF 1984
4 *In the beginning – the Manchester origins of Rolls-Royce*
Mike Evans, RRHT 2nd edition 2004
5 *Rolls-Royce – the Derby Bentleys*, Alec Harvey-Bailey, SHRMF 1985
6 *The early days of Rolls-Royce - and the Montagu family*
Lord Montagu of Beaulieu, RRHT 1986
7 *Rolls-Royce – Hives, the quiet tiger*, Alec Harvey-Bailey, SHRMF 1985
8 *Rolls-Royce – Twenty to Wraith*, Alec Harvey-Bailey, SHRMF 1986
9 *Rolls-Royce and the Mustang*, David Birch, RRHT 1997
10 *From Gipsy to Gem with diversions, 1926-1986*, Peter Stokes, RRHT 1987
11 *Armstrong Siddeley - the Parkside story, 1896-1939*
Ray Cook, RRHT 1989
12 *Henry Royce – mechanic*, Donald Bastow, RRHT 1989
14 *Rolls-Royce - the sons of Martha*, Alec Harvey-Bailey, SHRMF 1989
15 *Olympus - the first forty years*, Alan Baxter, RRHT 1990
16 *Rolls-Royce piston aero engines - a designer remembers*
A A Rubbra, RRHT 1990
17 *Charlie Rolls – pioneer aviator*, Gordon Bruce, RRHT 1990
18 *The Rolls-Royce Dart - pioneering turboprop*, Roy Heathcote, RRHT 1992
19 *The Merlin 100 series - the ultimate military development*
Alec Harvey-Bailey and Dave Piggott, RRHT 1993
20 *Rolls-Royce – Hives' turbulent barons*, Alec Harvey-Bailey, SHRMF 1992
21 *The Rolls-Royce Crecy*, Nahum, Foster-Pegg and Birch, RRHT 1994
22 *Vikings at Waterloo - the wartime work on the Whittle jet engine by the Rover Company*, David S Brooks, RRHT 1997
23 *Rolls-Royce - the first cars from Crewe*, K E Lea, RRHT 1997
24 *The Rolls-Royce Tyne*, L Haworth, RRHT 1998
25 *A View of Ansty*, D E Williams, RRHT 1998
26 *Fedden – the life of Sir Roy Fedden*, Bill Gunston, RRHT 1998
27 *Lord Northcliffe – and the early years of Rolls-Royce*
Hugh Driver, RREC 1998
28 *Boxkite to Jet – the remarkable career of Frank B Halford*
Douglas R Taylor, RRHT 1999
29 *Rolls-Royce on the front line – the life and times of a Service Engineer*
Tony Henniker, RRHT 2000
30 *The Rolls-Royce Tay engine and the BAC One-Eleven*
Ken Goddard, RRHT 2001
31 *An account of partnership – industry, government and the aero engine,*
G P Bulman, RRHT 2002
32 *The bombing of Rolls-Royce at Derby in two World Wars – with diversions,*
Kirk, Felix and Bartnik, RRHT 2002
33 *Early Russian jet engines – the Nene and Derwent in the Soviet Union, and the evolution of the VK-1*
Vladimir Kotelnikov and Tony Buttler, RRHT 2003
34 *Pistons to Blades, small gas turbine developments by the Rover Company,*
Mark C S Barnard, RRHT 2003
35 *The Rolls-Royce Meteor – Cromwell and other applications*
David Birch, RRHT 2004
36 *50 years with Rolls-Royce – my reminiscences*, Donald Eyre, RRHT 2005
37 *Stoneleigh Motors – an Armstrong Siddeley company,*
Alan Betts, RRHT 2006
38 *The Life and Times of Henry Edmonds*, Paul Tritton, RRHT 2006

Technical Series

1 *Rolls-Royce and the Rateau Patents*, H Pearson, RRHT 1989
2 *The vital spark! The development of aero engine sparking plugs*
K Gough, RRHT 1991
3 *The performance of a supercharged aero engine*
S Hooker, H Reed and A Yarker, RRHT 1997
4 *Flow matching of the stages of axial compressors*
Geoffrey Wilde OBE, RRHT 1999
5 *Fast jets – the history of reheat development at Derby*
Cyril Elliott, RRHT 2001
6 *Royce and the vibration damper*, T C Clarke, RRHT 2003
7 *Rocket development with liquid propellants*
W H J Riedel, RRHT 2005 (translated by John Kelly 2004)
8 *Fundamentals of car performance*
Hives, Lovesey, Robotham, RRHT 2006

Specials

* *Sectioned drawings of piston aero engines*, L Jones, RRHT 1995
* *Hall of Fame*, RRHT 2004
* *Rolls-Royce Centenary Luncheon*, M H Evans, 2005
* *Rolls-Royce Armaments*, D Birch, RRHT 2000